D0212773

FRICTION, WEAR, LUBRICATION

A Textbook in Tribology

Kenneth C Ludema
Professor of Mechanical Engineering
The University of Michigan
Ann Arbor, Michigan

CRC Press
Boca Raton New York London Tokyo

The cover background is a photograph of a steel surface (light blue) partially covered with streaks of "protective" film due to sliding in engine oil. The image was created by a polarizing interference (Françon) microscope objective (25×) with about 40× further magnification.

The graph on the front cover shows that the "protective" film builds up progressively and, if it functions quickly enough, it will prevent scuffing failure of the sliding surface.

Acquiring Editor:	Norm Stanton
Editorial Assistant	Jennifer Petralia
Project Editor:	Gail Renard
Marketing Manager:	Susie Carlisle
Cover design:	Denise Craig
PrePress:	Kevin Luong
Manufacturing:	Sheri Schwartz

Library of Congress Cataloging-in-Publication Data

Ludema, K.C
Friction, wear, lubrication : a textbook in tribology / by K.C Ludema.
p. cm.
Includes bibliographical references and index.
ISBN 0-8493-2685-0 (alk. paper)
1. Tribology.
TJ 1075.L84 1996
621.8′9—dc20 96-12440CIP

Direct all inquiries to CRC Press, Inc., 2000 Corporate Blvd., N.W., Boca Raton, Florida 33431.

International Standard Book Number 0-8493-2685-0
Library of Congress Card Number 96-12440
Printed in the United States of America 1 2 3 4 5 6 7 8 9 0
Printed on acid-free paper

ABOUT THE AUTHOR

Kenneth C Ludema is Professor of Mechanical Engineering at the University of Michigan in Ann Arbor. He holds a B.S. degree from Calvin College in Grand Rapids, Michigan, an M.S. and Ph.D. from the University of Michigan, and a Ph.D. from Cambridge University. He has been on the faculty of the University of Michigan since 1962 and has taught courses in materials, manufacturing processes, and tribology.

Dr. Ludema, along with his students, has published more than 75 papers, primarily on mechanisms of friction and boundary lubrication.

ABOUT THIS BOOK

This book is intended primarily to be used as a textbook, written on the level of senior and graduate students with proficiency in engineering or sciences. It is intended to bring everyone who wants to solve problems in friction and wear to the same understanding of what is (and, more important, what is not) involved. Most engineers and scientists have learned a few simple truths about friction and wear, few of which seem relevant when problems arise. It turns out that the "truths" are often too simple and couched too much in the terms of the academic discipline in which they have been taught. This book suggests a different approach, namely, to explore the tribological behavior of systems by well-designed experiments and tests, and to develop your own conclusions.

One useful way to control *friction* and *wear* is by *lubrication*, though it is often not the economical way. These three topics together constitute the broad area of tribology. Tribology has many entry points because of its great breadth. The advancement of each of its subtopics requires concentrated effort, and many people spend a satisfying and useful career in only one of them. By contrast, product designers and engineers need to be moderately proficient in all related topics with some understanding of the more specialized topics.

THE STATUS OF TRIBOLOGY

Tribology as a whole lags behind engineering in general in the development of equations, formulae, and methods for general use in engineering design. Indeed, there are some useful methods and equations available, mostly in full film fluid lubrication and contact stress calculations. The reason for the advanced state of these topics is that very few variables are needed to characterize adequately the system under study, namely, fluid properties and geometry in the subject of liquid lubrication, and elastic properties of solids and geometry in contact stress problems. A few more variables are required to estimate the temperature rise of sliding surfaces, but a great number are needed in useful equations for friction and wear.

The shortage of good design methods for achieving desired friction and product life virtually always results in postponing these considerations in product development until mere days before production. By this time the first choice for materials, processes, shapes, and part function is already locked in. The easy problems are solved first, such as product weight, strength, vibration characteristics, production methods, and cost. In the absence of formalized knowledge in friction and wear the engineering community resorts to guesswork, anecdotal information from vendors of various products including lubricants and materials, randomly selected accelerated tests done with totally inappropriate bench tests, and general over-design to achieve design goals. That need not be, and it has profound effects: the warranty costs for problems in friction and wear exceed the combined warranty costs for all other causes of product "failure" in the automotive and related industries.

LEARNING TRIBOLOGY

Tribology is ultimately an applied art and as such should be based upon, or requires background knowledge in, many topics. It is not a science by itself although research is done in several different sciences to understand the fundamental aspects of tribology. This, unfortunately, has had the effect of perpetuating (and even splintering) the field along disciplinary lines. One wit has expressed this problem in another sphere of life in the words, "England and America are divided by a common language." Often people from the various disciplines and the ever-present vendors offer widely different solutions to problems in tribology, which bewilders managers who would like to believe that tribology is a simple and straightforward art.

In academic preparation for designing products, most students in mechanical engineering (the seat of most design instruction) have taken courses in such topics as:

a. Fluid mechanics
b. Elasticity (described as solid mechanics)
c. Materials science (survey of atomic structure and the physics of solids)
d. Dynamics (mechanical mostly)
e. Heat transfer
f. Methods of mechanical design.

These are useful tools indeed, but hardly enough to solve a wide range of problems in friction and wear. Students in materials engineering will have a different set of tools and will gravitate toward those problems in which their proficiencies can be applied. But the complete tribologist will have added some knowledge in the following:

g. Plasticity
h. Visco-elasticity
i. Contact mechanics
j. The full range of mechanical properties of monolithic materials, composite materials, and layered structures (coatings, etc.), especially fracture toughness, creep, fatigue (elastic and low cycle)
k. Surface chemistry, oxidation, adhesion, adsorption
l. Surface-making processes
m. Statistical surface topographical characterization methods
n. Lubricant chemistry
o. and several more.

Many of these topics are addressed in this book, though it would be well for students to consult specialized books on these topics.

THE ORGANIZATION OF THE BOOK

Following are 14 chapters in which insight is offered for your use in solving tribological problems:

- *Chapter 1* informs you where to find further information on tribology and discusses the four major disciplines working in the field.

The next four chapters summarize some of the academic topics that may or should have been a part of the early training of tribologists:

- *Chapter 2* asserts that friction and wear resistance are separate from the usual mechanical properties of materials and cannot be adequately described in terms of those properties (though many authors disagree).
- *Chapter 3* discusses atomic structure, atomic energy states, and a few phenomena that are virtually always ignored in the continuum approach to modeling of the sliding process (and should not be).
- *Chapter 4* shows how real surfaces are made and discusses the inhomogeneous nature of the final product.
- *Chapter 5* is a short summary of the complicated topics of contact mechanics and temperature rise of sliding surfaces, in perspective.

Then, four chapters cover the core of tribology:

- *Chapter 6* gives a historical account of friction, presenting two major points:
  a. Causes for the great variability and unpredictability of friction, and
  b. What is required to measure friction reliably.
- *Chapter 7* is a synopsis of conventional lubrication — not much, but enough to understand its importance.
- *Chapter 8* discusses wear and provides an analysis of the many types and mechanisms seen in the technical literature. It discusses the actual events that cause loss of material from a sliding/rolling interface.
- *Chapter 9* is on chemical aspects of lubrication, where friction, wear, and lubrication converge in such problems as scuffing failure and break-in.

The following three chapters discuss methods of solving problems in friction and wear:

- *Chapter 10* is an analysis of design equations in friction and wear, showing that useful equations require more realistic assumptions than superposition of individual, steady state mechanisms of wear.
- *Chapter 11* suggests some useful steps in acquiring data on the friction and wear rates of components and materials for the design of mechanical components, both the technical and human aspects of the effort.
- *Chapter 12* describes how to diagnose wear problems and lists the attributes of the most common instruments for aiding analysis.

The last two chapters cover topics that could have been tucked into obscure corners of earlier chapters, but would have been lost there:

- *Chapter 13* is on coatings, listing some of the many types of coatings but showing that the nature of wear depends on the thickness of the coating relative to the size of the strain field that results from tribological interaction.
- *Chapter 14* covers bearings and materials, lightly.

A minimum of references has been used in this text since it is not primarily a review of the literature. In general, each chapter has a list of primary source books which can be used for historical perspective. Where there is no such book, detailed reference lists are provided.

There are problems sets for most of the chapters. Readers with training in mechanics will probably have difficulty with the problems in materials or physics; materialists will have trouble with mechanics; and scientists may require some time to fathom engineering methods. Stay with it! Real problems need all of these disciplines as well as people who are willing to gain experience in solving problems.

This book is the "final" form of a set of course notes I have used since 1964. Hundreds of students and practicing engineers have helped me over the years to gain my present perspective on the complicated and fascinating field of tribology.

I hope you will find this book to be useful.

Ken Ludema
Ann Arbor, Michigan
January 1996

CONTENTS

LIST OF TABLES

CHAPTER 1

The State of Knowledge in Tribology

TRIBOLOGY IS THE "OLOGY" OR SCIENCE OF "TRIBEIN." THE WORD COMES FROM THE SAME GREEK ROOT AS "TRIBULATION." A FAITHFUL TRANSLATION DEFINES TRIBOLOGY AS THE STUDY OF RUBBING OR SLIDING. THE MODERN AND BROADEST MEANING IS THE STUDY OF FRICTION, LUBRICATION, AND WEAR.

Tribological knowledge in written form is expanding at a considerable rate, but is mostly exchanged among researchers in the field. Relatively little is made available to design engineers, in college courses, in handbooks, or in the form of design algorithms, because the subject is complicated.

AVAILABLE LITERATURE IN TRIBOLOGY

Publishing activity in tribology is considerable, as is indicated by the number of papers and books published on the subject in one year. The main publications include the following:

Journals and Periodicals

Wear, published fortnightly by Elsevier Sequoia of Lausanne, Switzerland, produced 11 volumes in 1995 (180 through 190), containing 224 papers, and with indexes, editorials, etc., comprised 2752 pages. The papers are mostly on wear and erosion; some discuss contact mechanics; some deal with surface topography; and others are on lubrication, both liquid and solid.

Journal of Tribology (formerly the *Journal of Lubrication Technology*), one of the several Transactions of the ASME (American Society of Mechanical Engineers), published quarterly, produced Volume 116 in 1994 containing 109 papers, and with editorials, etc., comprised 876 pages. This journal is more mathematical than most others in the field, attracting papers in hydrodynamics, fluid rheology, and solid mechanics.

Tribology Transactions of the Society of Tribologists and Lubrication Engineers, or STLE, (formerly the American Society of Lubrication Engineers, or ASLE), published quarterly, produced Volume 37 in 1994, containing 113 papers, which together with miscellaneous items comprised 882 pages. The papers are mostly on lubricant chemistry and solid lubrication with some on hydrodynamics and scuffing. STLE also produces the monthly magazine, *Lubrication Engineer,* which contains some technical papers.

Tribology International is published bimonthly by IPC of London, and in 1993 produced Volume 25, containing 41 papers covering 454 pages, along with editorials, book reviews, news, and announcements. The papers cover a wide range of topics and are often thorough reviews of practical problems.

About 400 papers were published in Japanese journals, and many more in German, French, Russian, and Scandinavian journals. Some work is published in Chinese, but very little in Spanish, Portuguese, Hindi, or the languages of southern and eastern Europe, the middle east, or most of Africa.

In addition, there are probably 500 trade journals that carry occasional articles on some aspect of tribology. Some of these are journals in general design and manufacturing, and others are connected with such industries as those devoted to the making of tires, coatings, cutting tools, lubricants, bearings, mining, plastics, metals, magnetic media, and very many more. The majority of the articles in the trade journals are related to the life of a product or machine, and they only peripherally discuss the mechanisms of wear or the design of bearings. Altogether, over 10,000 articles are cited when a computer search of the literature is done, using a wide range of applicable key words.

Books

About 5 new books appear each year in the field, some of which may contain the word "tribology" in the title, while others may cover coatings, contact mechanics, lubricant chemistry, and other related topics.

There are several handbooks in tribology, of which the best known are:

- *The Wear Control Handbook of the ASME*, 1977 (Eds. W. Winer and M. Peterson).
- *The ASLE* (now STLE) *Handbook of Lubrication,* Vol. 1, 1978, Vol. 2, 1983, published by CRC Press (Ed. E.R. Booser).
- *The Tribology Handbook*, 1989, published by Halstead Press (Ed. M.J. Neale).
- *The ASM* (Vol. 18) *Handbook of Tribology,* 1994 (Ed. P.J. Blau).

Each of these handbooks has some strengths and some weaknesses. *The Tribology Handbook* is narrowly oriented to automotive bearings. The ASME *Wear Control Handbook* attempts to unify concepts across lubrication and wear through the simple Archard wear coefficients. The others contain great amounts of information, but that information is often not well coordinated among the many authors.

CONFERENCES ON FRICTION, LUBRICATION, AND WEAR HELD IN THE U.S.

Every year there are several conferences. Those of longest standing are the separate conferences of ASME and STLE and the joint ASME/STLE conference.

A separate, biannual conference, held in odd-numbered years in the U.S., is the Conference on Wear of Materials. The Proceedings papers are rigorously reviewed and until 1991 appeared in volumes published by the ASME. In 1993 the Proceedings became Volumes 163 through 165 of *Wear* journal, the 1995 proceedings became Volumes 181 through 183 (956 pages).

Another separate, biannual conference, held in even-numbered years, is the Gordon Conference on Tribology. It is a week-long conference held in June, at which about 30 talks are given but from which no papers are published.

Several *ad hoc* conferences are sponsored on some aspect of friction, lubrication, or wear by ASM, the American Society for Testing and Materials, the American Chemical Society, the Society of Plastics Engineers, the American Ceramic Society, the American Welding Society, the Society of Automotive Engineers, and several others.

THE SEVERAL DISCIPLINES IN THE FIELD OF TRIBOLOGY

Valiant attempts are under way to unify thinking in tribology. However, a number of philosophical divisions remain, and these persist in the papers and books being published. Ultimately, the divisions can be traced to the divisions in academic institutions. The four major ones are:

1. *Solid Mechanics*: focus is on the mathematics of contact stresses and surface temperatures due to sliding. Workers with this emphasis publish some very detailed models for the friction and wear rates of selected mechanical devices that are based on very simple physical tribological mechanisms.
2. *Fluid Mechanics*: focus is on the mathematics of liquid lubricant behavior for various shapes of sliding surfaces. Work in this area is the most advanced of all efforts to model events in the sliding interface for cases of thick films relative to the roughness of surfaces. Some work is also done on the influence of temperature, solid surface roughness, and fluid rheology on fluid film thickness and viscous drag. However, efforts to extend the methods of fluid mechanics to boundary lubrication are not progressing very well.
3. *Material Science*: focus is on the atomic and microscale mechanisms whereby solid surface degradation or alteration occurs during sliding. Work in this area is usually presented in the form of micrographs, as well as energy spectra for electrons and x-rays from worn surfaces. Virtually all materials, in most states, have been studied. Little convergence of conclusions is evident at this time, probably for two reasons. First, the limit of knowledge in the materials aspects of tribology has not yet been found. Second, material scientists (engineers,

physicists) rarely have a broad perspective of practical tribology. (Materials engineers often prefer to be identified as experts in wear rather than as tribologists.)

4. *Chemistry*: focus is on the reactivity between lubricants and solid surfaces. Work in this area progresses largely by orderly chemical alteration of bulk lubricants and testing of the lubricants with bench testers. The major deficiency in this branch of tribology is the paucity of work on the chemistry in the contacting and sliding conjunction region.

Work in each of these four areas is very detailed and thorough, and each requires years of academic preparation. The deficiencies and criticisms implied in the above paragraphs should not be taken personally, but rather as expressions of unmet needs that lie adjacent to each of the major divisions of tribology. There is little likelihood of any person becoming expert in two or even three of these areas. The best that can be done is for interdisciplinary teams to be formed around practical problems. Academic programs in general tribology may appear in the future, which may cut across the major disciplines given above. They are not available yet.

THE CONSEQUENCES OF FRICTION AND WEAR

The consequences of friction and wear are many. An arbitrary division into five categories follows, and these are neither mutually exclusive nor totally inclusive.

1. *Friction and wear usually cost money*, in the form of energy loss and material loss, as well as in the social system using the mechanical devices.

An interesting economic calculation was made by Jacob Rowe of London in 1734. He advertised an invention which reduces the friction of shafts. In essence, the main axle shaft of a wagon rides on two disks that have their own axle shaft. Presumably a saving is experienced by turning the second shaft more slowly than the wheel axle. Rowe's advertisement claimed: "All sorts of wheel carriage improved... a much less than usual draught of horses, etc., will be required in wagons, carts, coaches, and all other wheel vehicles as likewise all water mills, windmills, and horse mills... An estimate of the advantages that will accrue to the public, by means of canceling the friction of the wheel, pulley, balance, pendulum, etc..." (He then calculates that 40,000 horses are employed in the kingdom in wheel carriage, which number could be reduced to 20,000 because of the 2 to 1 advantage of his invention. At a cost of 15.5 shillings per day, the saving amounts to £1,095,000 per annum or £3000 per day.) In one sense, this would appear to reduce the number of horses needed, but Rowe goes on to say with enthusiasm that "great numbers of mines will be worked more than at present, and such as were not practicable before because of their remote distance from water and the poorness of the ore (so the carriage to the mills of water... eats up the profit) will now be carried on wheel carriages at a vastly cheaper rate than hitherto, and consequently there will be a greater demand for horses than at present, only, I must own that there will not be occasion to employ so large and heavy horses as common, for the draught that is now required being considerably less than usual

> shall want horses for speed more than draught." Another advantage of this new bearing, said Rowe, is that it will be far easier to carry fertilizer "and all sorts of dressing for lands so much cheaper than ordinary... great quantities of barren land will now be made fertile, which the great charges by the common way of carriage has hitherto rendered impracticable."

As to wear, it has been estimated by various agencies and committees around the world that wear costs each person between \$25 and \$250 per year (in 1966) depending upon what is defined as wear.[1] There are direct manifestations of wear, such as the wearing out of clothing, tires, shoes, watches, etc. which individually we might calculate easily. The cost of wear on highways, delivery trucks, airplanes, snowplows, and tree trimmers is more difficult to apply accurately to each individual. For the latter, we could take the total value of items produced each year on the assumption that the items produced replace worn items. However, in an expanding economy or technology new items become available that have not existed before, resulting in individuals accumulating goods faster than the goods can be worn out. Style changes and personal dissatisfaction with old items are also reasons for disposal of items before they are worn out.

An indirect cost in energy may be seen in automobiles, which are often scrapped because only a few of their parts are badly worn. Since the manufacture of an automobile requires as much energy as is required to operate that automobile for 100,000 miles, extending the life of the automobile saves energy.

2. *Friction and wear can decrease national productivity.* This may occur in several ways. First, if American products are less desirable than foreign products because they wear faster, our overseas markets will decline and more foreign products will be imported. Thus fewer people can be employed to make these products. Second, if products wear or break down very often, many people will be engaged in repairing the items instead of contributing to national productivity. A more insidious form of decrease in productivity comes about from the declining function of wearing devices. For example, worn tracks on track-tractors (bulldozers) cause the machine to be less useful for steep slopes and short turns. Thus, the function of the machine is diminished and the ability to carry out a mission is reduced. As another example, worn machine tool ways require a more skilled machinist to operate than do new machines.

3. *Friction and wear can affect national security.* The down time or decreased efficiency of military hardware decreases the ability to perform a military mission. Wear of aircraft engines and the barrels of large guns are obvious examples. A less obvious problem is the noise emitted by worn bearings and gears in ships, which is easily detectable by enemy listening equipment. Finally, it is a matter of history that the development of high-speed cutting tool steel in the 1930s aided considerably in our winning World War II.

4. *Friction and wear can affect quality of life.* Tooth fillings, artificial teeth, artificial skeletal joints, and artificial heart valves improve the quality of life when natural parts wear out. The wear of "external" materials also decreases the quality of life for many. Worn cars rattle, worn zippers cause uneasiness, worn watches

make you late, worn razors leave "nubs," and worn tires require lower driving speeds on wet roads.

5. *Wear causes accidents*. Traffic accidents are sometimes caused by worn brakes or other worn parts. Worn electrical wiring and switches expose people to electrical shock; worn cables snap; and worn drill bits cause excesses which often result in injury.

THE SCOPE OF TRIBOLOGY

Progress may be seen by contrasting automobile care in 1996 with that for earlier years. The owner's manual for a 1916 Maxwell automobile lists vital steps for keeping their deluxe model going, including:

Lubrication

Every day or every 100 miles

- Check oil level in the engine, oil lubricated clutch, transmission, and differential gear housing
- Turn grease cup caps on the 8 spring bolts, one turn (≈0.05 cu.in.)
- Apply a few drops of engine oil to steering knuckles
- Apply a few drops of engine oil to tie rod clevises
- Apply a few drops of engine oil to the fan hub
- Turn the grease cup on the fan support, one turn

Each week or 500 miles

- Apply a few drops of engine oil to the spark and throttle cross-shaft brackets
- Apply sufficient amounts of engine oil to all brake clevises, oilers, and cross-shaft brackets, at least 12 locations
- Force a "grease gun full" (half cup) of grease into the universal joint
- Apply sufficient engine oil to the starter shaft and switch rods
- Apply a few drops of engine oil to starter motor front bearing
- Apply a few drops of engine oil to the steering column oiler
- Turn the grease cup on the generator drive shaft, one turn
- Turn the grease cup on the drive shaft bearing, one turn
- Pack the ball joints of the steering mechanism with grease (≈ 1/4 cup)
- Apply a few drops of engine oil to the speedometer parts

Each month or 1500 miles

- Force a "grease gun full" of grease into the engine timing gear
- Force a "grease gun full" of grease into the steering gear case
- Apply a few drops of 3-in-1 oil to the magneto bearings
- Pack the wheel hubs with grease (≈ 1/4 cup each)
- Turn the grease cup on the rear axle spring seat, two turns

Each 2000 miles

- Drain crank case, flush with kerosene, and refill (several quarts)
- Drain wet clutch case, flush with kerosene, and refill (≈ one quart)
- Drain transmission, flush with kerosene, and refill (several quarts)

- Drain rear axle, flush with kerosene, and refill (≈ 2 quarts)
- Jack up car by the frame, pry spring leaves apart, and insert graphite grease between the leaves

Other Maintenance

Every two weeks	**On a regular basis**
Check engine compression	Check engine valve action
Listen for crankshaft bearing noises	Inspect ignition wiring
Clean and regap spark plugs	Check battery fluid level and color
Adjust carburetor mixtures	Inspect cooling system for leaks
Clean gasoline strainer	Check fan belt tension
Drain water from carburetor bowl	Inspect steering parts
Inspect springs	Tighten body and fender bolts
Check strength of magneto spark	Check effectiveness of brakes
Check for spark knock, to determine when carbon should be removed from head of engine	Examine tires for cuts or bruises
	Adjust alcohol/water ratio in radiator

If an automobile of that era survived 25,000 miles it was uncommon, partly because of poor roads but also because of high wear rates. The early cars polluted the streets with oil and grease that leaked though the seals. The engine burned a quart of oil in less than 250 miles when in good condition and was sometimes not serviced until an embarrassing cloud of smoke followed the car. Fortunately there were not many of them! Private garages of that day had dirt floors, and between the wheel tracks the floor was built up several inches by dirt soaked with leaking oil and grease. We have come a long way.

Progress since the 1916 Maxwell has come about through efforts in many disciplines:

1. Lubricants are more uniform in viscosity, with harmful chemical constituents removed and beneficial ones added
2. Fuels are now carefully formulated to prevent pre-ignition, clogging of orifices in the fuel system, and excessive evaporation
3. Bearing materials can better withstand momentary loss of lubricant and overload
4. Manufacturing tolerances are much better controlled to produce much more uniform products, with good surface finish
5. The processing of all materials has improved to produce homogeneous products and a wider range of materials, metals, polymers, and ceramics
6. Shaft seals have improved considerably

Progress has been made on all fronts, but not simultaneously. The consumer product industry tends to respond primarily to the urgent problems of the day, leaving others to arise as they will. However, even when problems in tribology arise they are more often seen as vexations rather than challenges.

REFERENCE

1. *H.P. Jost Reports,* Committee on Tribology, Ministry of Technology and Industry, London, 1966.

CHAPTER 2

Strength and Deformation Properties of Solids

WEAR LIFE EQUATIONS USUALLY INCLUDE SYMBOLS THAT REPRESENT MATERIAL PROPERTIES. WITH FEW EXCEPTIONS THE MATERIAL PROPERTIES ARE THOSE THAT REFLECT ASSUMPTIONS OF ONE OR TWO MATERIALS FAILURE MODES IN THE WEARING PROCESS. IT WILL BE SHOWN LATER IN THE BOOK THAT THE WEAR RESISTING PROPERTIES OF SOLIDS CANNOT GENERALLY BE DESCRIBED IN TERMS OF THEIR MECHANICAL PROPERTIES JUST AS ONE MECHANICAL PROPERTY (E.G., HARDNESS) CANNOT BE CALCULATED FROM ANOTHER (E.G., YOUNG'S MODULUS).

INTRODUCTION

Sliders, rolling contacters, and eroding particles each impose potentially detrimental conditions upon the surface of another body, whether the scale of events is macroscopic or microscopic. The effects include strains, heating, and alteration of chemical reactivity, each of which can act separately but each also alters the rate of change of the others during continued contact between two bodies.

The focus in this chapter is upon the strains, but expressed mostly in terms of the stresses that produce the strains. Those stresses, when of sufficient magnitude and when imposed often enough upon small regions of a solid surface, will cause fracture and eventual loss of material. It might be expected therefore that equations and models for wear rate should include variables that relate to imposed stress and variables that relate to the resistance of the materials to the imposed stress. These latter, material properties, include Young's Modulus (E), stress intensity factor (K_c), hardness (H), yield strength (Y), tensile strength (S_u), strain to failure (ε_f), work hardening coefficient (n), fatigue strengths, cumulative variables in ductile fatigue, and many more. Though many wear equations have been published which incorporate material properties, none is widely applicable. The reason is that:

The stress states in tests for each of the material properties are very different from each other, and different again from the tribological stress states.

The importance of these differences will be shown in the following paragraphs and summarized in the section titled *Application to Tribology,* later in this chapter.

TENSILE TESTING

In elementary mechanics one is introduced to tensile testing of materials. In these tests the materials behave elastically when small stresses are applied. Materials do not actually behave in a linear manner in the elastic range, but linearly enough to base a vast superstructure of elastic deflection equations on that assumption. Deviations from linearity produce a hysteresis, damping loss, or energy loss loop in the stress–strain data such that a few percent of the input energy is lost in each cycle of strain. The most obvious manifestation of this energy loss is heating of the strained material, but also with each cycle of strain some damage is occurring within the material on an atomic scale.

As load and stress are increased, the elastic range may end in one of two ways, either by immediate fracture or by various amounts of plastic flow before fracture. In the first case, the material is considered to be brittle, although careful observation shows that no material is perfectly brittle. Figure 2.1a shows the stress–strain curve for a material with little ductility, i.e., a fairly brittle material. When plastic deformation begins, the shape of the stress–strain curve changes considerably. Figure 2.1b shows a very ductile material.

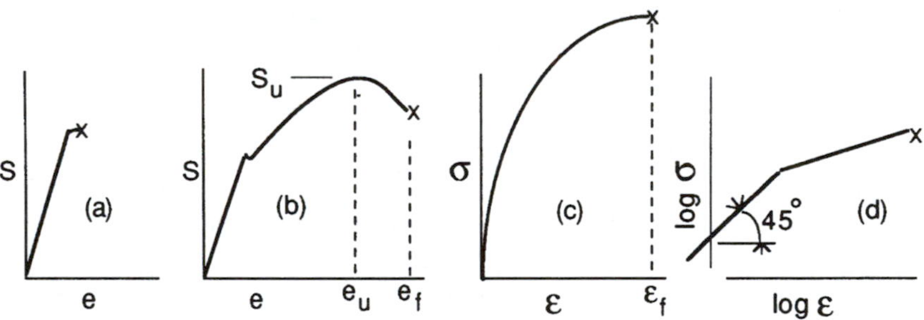

Figure 2.1 Stress–strain curves (x=fracture point).

In Figures 2.1a and 2.1b the ordinate, S, is defined as,

$$S = \frac{\text{applied load}}{\text{original cross-sectional area of the specimen}}$$

Plotting of stress by this definition shows an apparent weakening of material beyond the value of e where S is maximum, referred to as S_u. S_u is also referred to as the tensile strength (TS) of the material, but should rather be called the maximum load-carrying capacity of the tensile specimen. At that point the tensile specimen begins to “neck down” in one small region.

In Figures 2.1a and 2.1b the abscissa, e, is defined as:

$$e = \frac{\text{change in length of a chosen section of a tensile specimen}}{\text{original length of that section}}$$

The end point of the test is given as the % elongation property, which is $100e_f$.

Figure 2.1c is a stress–strain curve in which the ordinate is the true stress, σ, defined as:

$$\sigma = \frac{\text{applied load}}{\text{cross-sectional area, measured when applied load is recorded}}$$

The abscissa is the true strain, ε, defined as,

$$\varepsilon = \ln (A_1/A_2)$$

where A is the cross-sectional area of the tensile specimen, and measurement #2 was taken after measurement #1. Further, $\varepsilon = \ln(1 + e)$ where there is uniform strain, i.e., in regions far from the location of necking down.

The best-fit equation for this entire elastic-plastic curve is of the form $\sigma = K\varepsilon^n$. Figure 2.1c shows the true strength of the material, but obscures the load-carrying capacity of the tensile specimen. An interesting consequence of the necking down coinciding with the point of maximum load-carrying capacity is that $\varepsilon_u = n$.

Figure 2.1d shows the same data as given in Figure 2.1c, except on a log–log scale. The elastic curve is (artificially) constrained to be linear, and the data taken from tests in the plastic range of deformation plot as a straight line with slope "n" beyond $\varepsilon \approx 0.005$, i.e., well beyond yielding. The equation for this straight line (beyond $\varepsilon \approx 0.005$) is (again!) found to be $\sigma = K\varepsilon^n$. The representation of tensile data as given in Figure 2.1d is convenient for data reduction and for solving some problems in large strain plastic flow. The major problem with the representation of Figure 2.1d is that the yield point cannot be taken as the intersection of the elastic and plastic curves. For most metals, the yield point may be as low as two thirds the intersection, whereas for steel it is often above.

Tensile data are instructive and among the easiest material property data to obtain with reasonable accuracy. However, few materials are used in a state of pure uni-axial tension. Usually, materials have multiple stresses on them, both normal stresses and shear stresses. These stresses are represented in the three orthogonal coordinate directions as, x, y, and z, or 1, 2, and 3. It is useful to know what combination of three-dimensional stresses, normal and shear stresses, cause yielding or brittle failure. There is no theoretical way to determine the conditions for either mode of departure from elastic strain (yielding or brittle fracture), but several theories of "failure criteria" have been developed over the last two centuries.

(ELASTIC) FAILURE CRITERIA

The simplest of these failure criteria states that whenever a critical value of normal strain or normal stress, tensile or compressive, is applied in any direction, failure will occur. These criteria are not very realistic. Griffith (see reference number 4) and others found that in tension a brittle material fractures at a stress, σ_t, whereas a compression test of the same material will show that the stress at fracture is about $-8\sigma_t$. From these data Griffith developed a fracture envelope, called a fracture criterion, for brittle material with two-dimensional normal applied stresses, which may be plotted as shown in Figure 2.2.

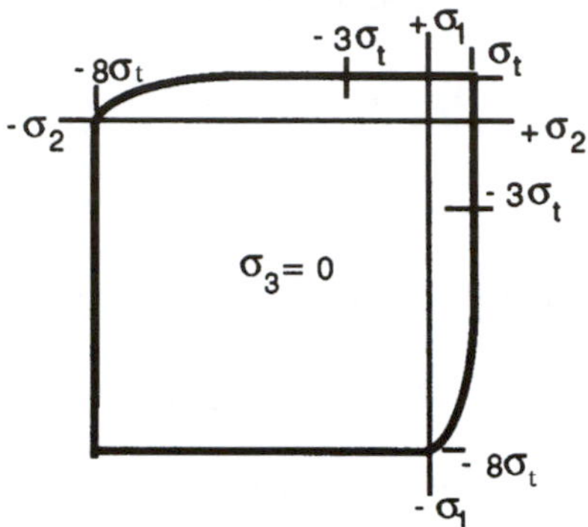

Figure 2.2 Graphical representation of the Griffith criterion for brittle fracture in biaxial normal stress.

PLASTIC FAILURE (YIELD CRITERIA)

There are also several yield criteria, as may be seen in textbooks on mechanics. One that is easily understood intuitively is the maximum shear stress theory, but one of the most widely used mathematical expression is that of von Mises,

$$(\sigma_x - \sigma_y)^2 + (\sigma_y - \sigma_z)^2 + (\sigma_z - \sigma_x)^2 + 6(\tau_{xy}{}^2 + \tau_{yz}{}^2 + \tau_{zx}{}^2) = 2Y^2 \qquad (1)$$

Y is the stress at which yielding begins in a tensile test, σ is the normal stress, and τ is the shear stress as shown in Figure 2.3. The von Mises equation states that any stress combination can be imposed upon an element of material, tensile (+), compressive (–), and shear, and the material will remain elastic until the proper summation of all stresses equals $2Y^2$. Note that the signs on the shear stresses have no influence upon the results.

The above two criteria, the Griffith criterion and the von Mises criterion, refer to different end results. The Griffith criterion states that *brittle* fracture results from tensile (normal) stresses predominantly, although compressive stresses impose shear stresses which also produce brittle failure. The von Mises criterion

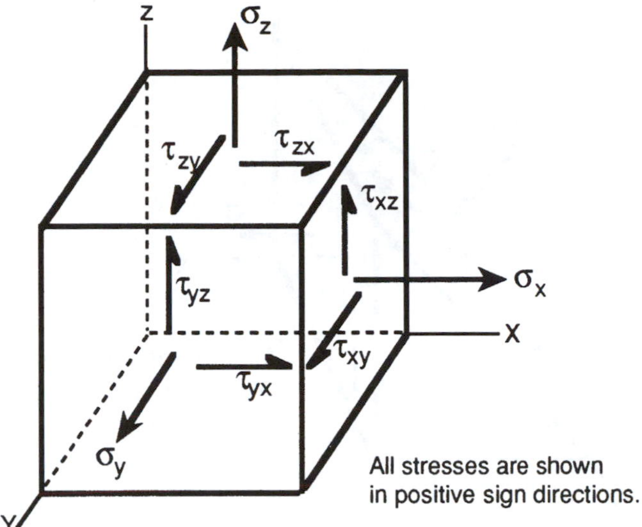

Figure 2.3 Stresses on a *point* assumed here to be constant over the cube faces.

states that combinations of *all* normal and shear stresses *together* result in *plastic* shearing. It is instructive to show the relationship between imposed stresses and the two modes of departure from elasticity, i.e., plastic flow and brittle cleavage. This begins with an exercise in transformation of axes of stress.

TRANSFORMATION OF STRESS AXES AND MOHR CIRCLES

A solid cube with normal and shear stresses imposed upon its faces can be cut as shown in Figure 2.4. The stresses σ_x and τ_{xz} imposed upon the x face (to the right) multiplied by the area of the x face constitutes applied forces on the x face, and likewise for the z face (at the bottom). The stresses $\sigma_{x'}$ and $\tau_{x'z'}$ on the x′ (slanted) face multiplied over the area of the x′ face constitute a force that must balance the two previous forces.

The stresses are related by the following equations:

$$\sigma_{x'} = \sigma_z \cos^2\alpha + \sigma_x \sin^2\alpha - 2\tau_{xz}\sin\alpha\ \cos\alpha$$

$$\tau_{x'z'} = (\sigma_z - \sigma_x)\sin\alpha\ \cos\alpha + \tau_{xz}(\cos^2\alpha - \sin^2\alpha) \qquad \textbf{(2)}$$

Equations can be written for wedges of orientations other than α. For example, on a plane oriented at α + 90° we would calculate the normal stress to be:

$$\sigma_{z'} = \sigma_z \sin^2\alpha + \sigma_x \cos^2\alpha - 2\tau_{xz}\sin\alpha\ \cos\alpha$$

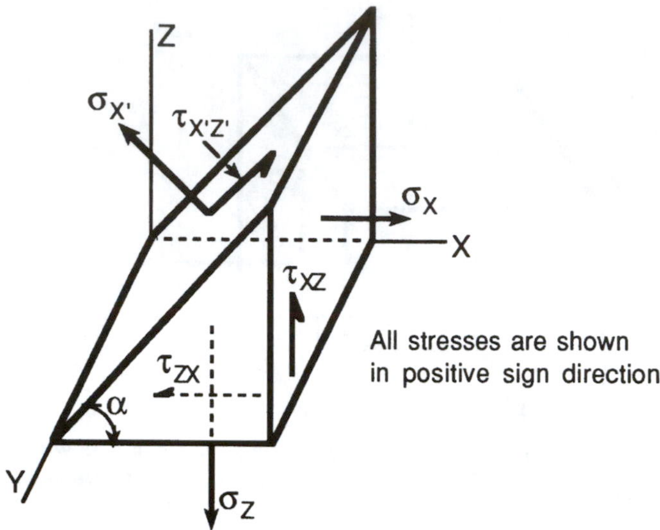

Figure 2.4 Stresses on the face of a wedge oriented at an angle α.

Otto Mohr developed a way to visualize the stresses on all possible planes (i.e., all possible values of α) by converting Equations 2 to double angles as follows:

$$\sigma_{z'} = \frac{(\sigma_z + \sigma_x)}{2} + \frac{(\sigma_z - \sigma_x)}{2}\cos 2\alpha + \tau_{xz}\sin 2\alpha$$

$$\sigma_{x'} = \frac{(\sigma_x + \sigma_z)}{2} + \frac{(\sigma_x - \sigma_z)}{2}\cos 2\alpha - \tau_{xz}\sin 2\alpha$$

$$\tau_{x'z'} = \frac{-(\sigma_x - \sigma_z)}{2}\sin 2\alpha + \tau_{xz}\cos 2\alpha \quad \textbf{(3)}$$

He plotted these equations upon coordinate axes in $\pm\,\sigma$ and $\pm\,\tau$ as shown in Figure 2.5. The values of $\sigma_{x'}$, $\sigma_{z'}$, and $\tau_{x'z'}$ for all possible values of α describe a circle on those axes. Two states of stress will now be shown on the Mohr axes, namely for a tensile test and for a torsion test.

In Figure 2.6 the tensile load is applied in the x direction and thus there is a finite stress σ_x in that direction. There is no applied normal stress in the y or z direction, nor shear applied in any direction: so $\sigma_y = \sigma_z = \tau_{xy} = \tau_{yz} = \tau_{zx} = 0$.

The state of stress on planes chosen at any desired angle relative to the applied load in a tensile test constitutes a circle on the Mohr axes as shown in Figure 2.5a. Points σ_x and 0 are located and a circle is drawn through these points around a center at $\sigma_x/2$. The normal and shear stresses on a plane oriented 45° from the

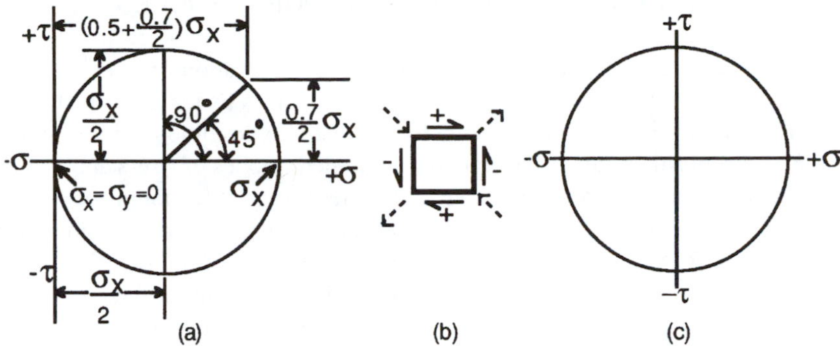

Figure 2.5 Mohr circles for tension (a) and torsion (c).

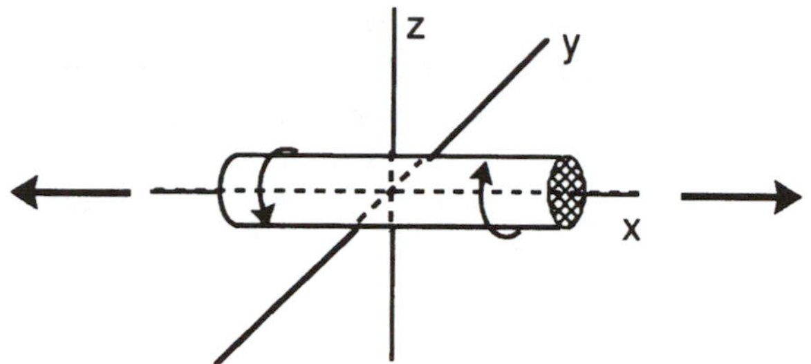

Figure 2.6 Orientation of test specimen with respect to a coordinate axis and positive direction of applied load torque.

x axis of the bar in Figure 2.5 are shown by drawing a line through the center of the Mohr circle and set at an angle of 90° (45° × 2) from the stress in the x direction. The normal stress and the shear stress on that plane in the specimen are both of magnitude $\sigma_x/2$. This can be verified by setting α=45° in Equation 2 or 3.

The stress state on any other plane can as easily be determined. For example, the stress state on a plane oriented 22.5° from the x direction in the specimen is shown by drawing a line from the center of the circle and set at an angle of 45° (22.5×2) from the applied stress in the x direction. The normal stress on that plane has the magnitude $\sigma_x/2 + (\sqrt{2}\ \sigma_x)/2$ and the shear stress is $(\sqrt{2}\ \sigma_x)/2$, as shown in Figure 2.5a.

The stress state upon an element in the surface of a bar in torsion is shown in Figure 2.5b. A set of balancing shear stresses comprises a *plus* shear stress and a *minus* shear stress. These stresses are shown on Mohr axes in Figure 2.5c. Note that these shear stresses can be resolved into a tensile stress and a compressive stress oriented 45° from the direction of the shear stresses. The directions of these stresses relative to the applied shear stresses are also shown in Figure 2.5b.

(See Problem Set questions 2 a, b, and c.)

MATERIAL PROPERTIES AND MOHR CIRCLES

One very useful feature of the Mohr circle representation of stress states is that *material properties* may be drawn on the same axes as the applied stresses, allowing a visualization of progression toward the two possible modes of departure from the elastic state via different (or combined) modes of stress application. These two are plastic (ductile) shearing and tensile (brittle) failure, two very different and independent properties of solid matter and worthy of some emphasis. (See Chapter 3, the section titled *Dislocations, Plastic Flow, and Cleavage*). These properties are not related, and are not connected with the common assumption that the shear strength of a material is half the tensile strength.

We will use a simple, straight-line representation of these properties, bypassing other (and perhaps more accurate) concepts under discussion in mechanics research. Our first example will be cast iron, which is generally taken to be a brittle material when tensile stresses are applied. Figure 2.7 shows a set of four circles for increasing applied tensile stress, with the shear strength and brittle fracture limits also shown. The critical point is reached when the circle touches the brittle fracture strength line, and the material fails in a brittle manner. This is observed in practice, and there can be few explanations other than that the shear strength of the cast iron is greater than half the brittle fracture strength, i.e., $\tau_y > \sigma_b/2$.

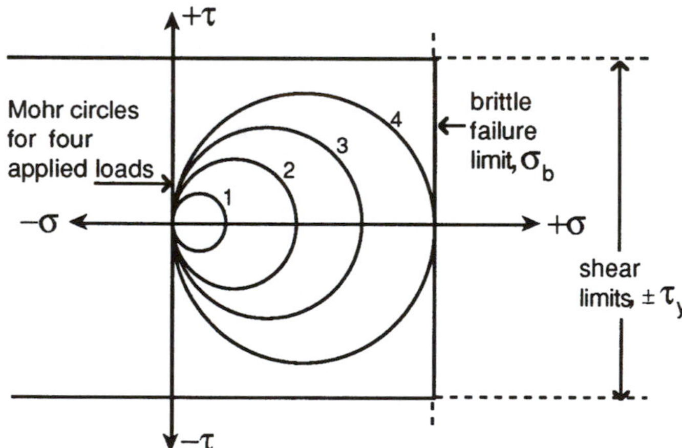

Figure 2.7 Mohr circles for tensile stresses in cast iron, ending in brittle fracture.

Figure 2.8 shows a set of circles for increasing torsion on a bar of cast iron. In this case the first critical point occurs when the third circle touches the (initial) shear strength line. This occurs because $\sigma_b > \tau_y$. The material plastically deforms as is observed in practice. With further strain the material work hardens, which may be shown by an increasing shear limit. Finally, the circle expands to touch the cleavage or brittle fracture strength of the material, and the bar fractures. Cast iron is thus seen to be a fairly ductile material in torsion. A half-inch-diameter bar of cast iron, six inches in length, may be twisted more than three complete revolutions before it fractures.

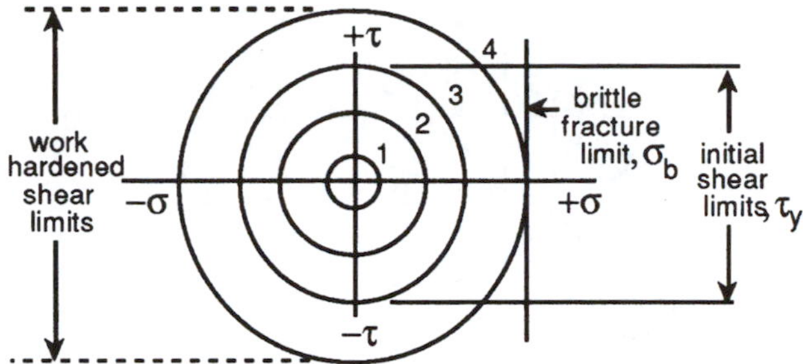

Figure 2.8 Mohr circles for increasing torsion on a bar of cast iron. The first "failure" occurs in plastic shear, followed by work hardening and eventual brittle failure.

This same type of exercise may be carried out with two other classes of material, namely, ductile metals and common ceramic materials. Ductile metals (partly by definition), always plastically deform before they fracture in either tension or torsion. Thus $\sigma_b > 2\tau_y$. Ceramic materials usually fail in a brittle manner in both tension and torsion (just as glass and chalk sticks do) so that $\sigma_b < \tau_y$.

(See Problem Set questions 2 d and e.)

VON MISES VERSUS MOHR (TRESCA) YIELD CRITERIA

So far only Mohr circles for tension or torsion (shear), separately, have been shown. In the more practical world the stress state on an element (cube) includes some shear stresses. If one face (of a cube) can be found with relatively little shear stress imposed, this shear stress can be taken as zero and a Mohr circle can be drawn. If all three coordinate directions have significant shear stresses imposed, it is necessary to use a cubic equation for the general state of stress at a point to solve the problem: these equations can be found in textbooks on solid mechanics.

If one face of a cube (e.g., the z face) has no shear stresses, that face may be referred to as a principal stress face. The other faces are assumed to have shear stresses τ_{xy} and τ_{yx} imposed. The Mohr circle can be constructed by looking into the z face first to visualize the stresses upon the other faces. The other stresses can be plotted as shown in Figure 2.9. Here σ_x is arbitrarily taken to be a small compressive stress and σ_y a larger tensile stress. The Mohr circle is drawn through the vector sum of σ and τ on each of the x and y faces. Again, the stresses on all possible planes perpendicular to the z face are shown by rotation around the origin of the circle. One interesting set of stresses is seen at angle θ (in the figure) from the stress states imposed upon the x and y faces. These are referred to as principal stresses, designated as σ_1 and σ_3, because of the absence of shear stress on these planes. (σ_2 is defined later.) These stresses may also be calculated by using the following equations:

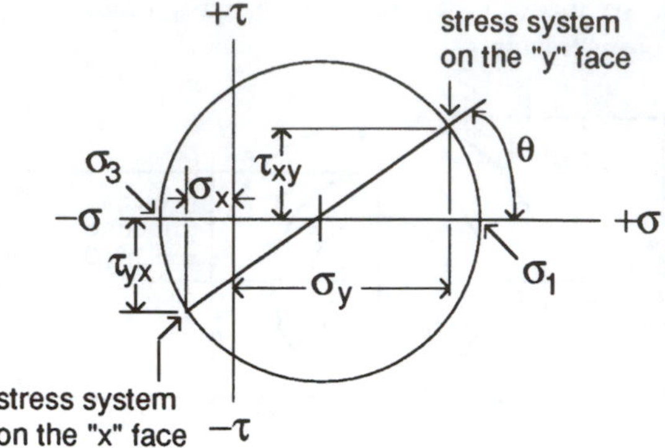

Figure 2.9 The Mohr circle for nonprincipal orthogonal stresses.

$$\sigma_1 = \sigma_{x'} = \frac{(\sigma_x + \sigma_y)}{2} + \sqrt{\frac{(\sigma_x - \sigma_y)^2}{4} + \tau_{xy}^2}$$

$$\sigma_3 = \sigma_{y'} = \frac{(\sigma_x + \sigma_y)}{2} - \sqrt{\frac{(\sigma_x - \sigma_y)^2}{4} + \tau_{xy}^2} \tag{4}$$

The principal stresses can be thought of as being imposed upon the surfaces of a new cube rotated relative to the original cube by an angle θ/2, as shown in Figure 2.10.

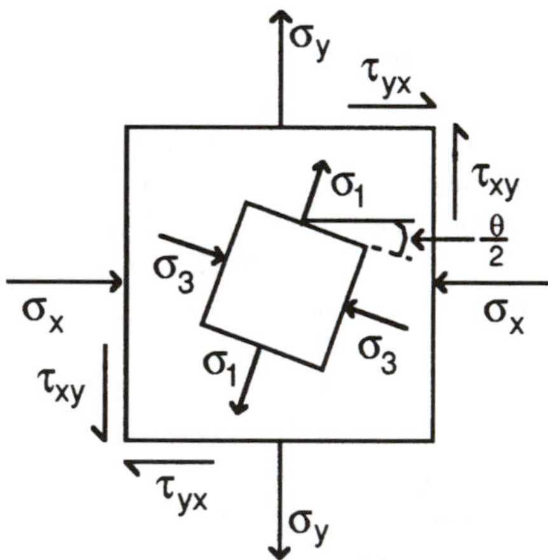

Figure 2.10 Resolving of nonprincipal stress state to a principal stress state (where there is no shear stress in the "2," i.e., z face).

Now that one circle is found, two more can be found by looking into the "1" and "3" faces. If σ_z is a tensile stress state of smaller magnitude than σ_y then it lies between σ_1 and σ_3 and is designated σ_2. By looking into the 1 face, σ_2 and σ_3 are seen, the circle for which is shown in Figure 2.11 as circle 1.

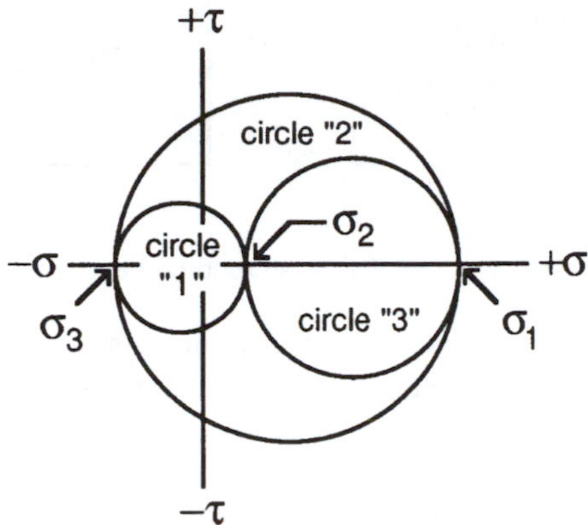

Figure 2.11 The three Mohr circles for a cube with only principal stresses applied.

Circle 2 is drawn in the same way. (Recall that in Figures 2.5, 2.7, and 2.8 only principal stresses were imposed.) The inner cube in Figure 2.10 has only principal stresses on it. In Figure 2.11 only those principal stresses connected with the largest circle contribute to yielding. The von Mises equation, Equation 1, suggests otherwise. (The Mohr circle embodies the Tresca yield criterion, incidentally.) Equation 1 for principal stresses only is:

$$(\sigma_1 - \sigma_2)^2 + (\sigma_2 - \sigma_3)^2 + (\sigma_3 - \sigma_1)^2 = 2Y^2 \qquad \textbf{(5)}$$

which can be used to show that the Tresca and von Mises yield criteria are identical when σ_2 = either σ_1 or σ_3, and farthest apart (≈15%) when σ_2 lies half way between. Experiments in yield criteria often show data lying between the Tresca and von Mises yield criteria.

VISCO-ELASTICITY, CREEP, AND STRESS RELAXATION

Polymers are visco-elastic, i.e., mechanically they appear to be elastic under high strain rates and viscous under low strain rates. This behavior is sometimes modeled by arrays of springs and dashpots, though no one has ever seen them in real polymers. Two simple tests show visco-elastic behavior, and a particular mechanical model is usually associated with each test, as shown in Figure 2.12. From these data of ε and σ versus time, it can be seen that the Young's Modulus, E ($=\sigma/\varepsilon$), decreases with time.

The decrease in E of polymers over time of loading is very different from the behavior of metals. When testing metals, the loading rate or the strain rates

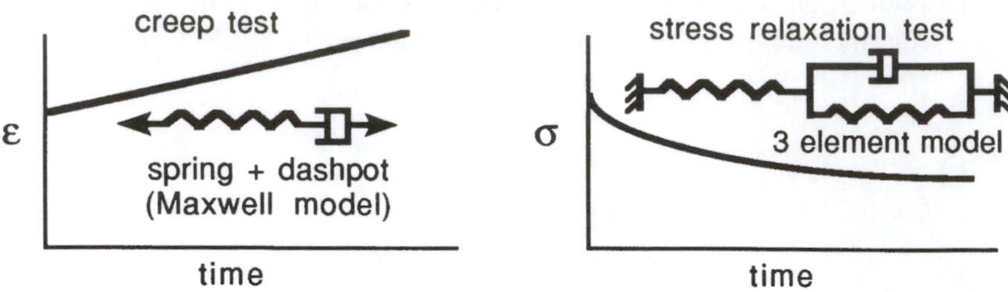

Figure 2.12 Spring/dashpot models in a creep test and a stress relaxation test.

are usually not carefully controlled, and accurate data are often taken by stopping the test for a moment to take measurements. That would be equivalent to a stress relaxation test, though very little relaxation occurs in the metal in a short time (a few hours).

For polymers which relax with time, one must choose a time after quick loading and stopping, at which the measurements will be taken. Typically these times are 10 seconds or 30 seconds. The 10-second values for E for four polymers are given in Table 2.1.

Table 2.1 Young's Modulus for Various Materials

Solid	E. Young's Modulus	
polyethylene	≈ 34,285 psi	(10s modulus)
polystyrene	≈ 485,700 psi	(10s modulus)
polymethyl-methacrylate	≈ 529,000 psi	(10s modulus)
Nylon 6-6	≈ 285,700 psi	(10s modulus)
steel	≈ 30×10^6 psi	(207 GPa)
brass	≈ 18×10^6 psi	(126 GPa)
lime-soda glass	≈ 10×10^6 psi	(69.5 GPa)
aluminum	≈ 10×10^6 psi	(69.5 GPa)

Dynamic test data are more interesting and more common than data from creep or stress relaxation tests. The measured mechanical properties are Young's Modulus in tension, E, or in shear, G, (strictly, the tangent moduli E' and G') and the damping loss (fraction of energy lost per cycle of straining), Δ, of the material. (Some authors define damping loss in terms of tan δ, which is the ratio E''/E' where E'' is the loss modulus.) Both are strain rate (frequency, f, for a constant amplitude) and temperature (T) dependent, as shown in Figure 2.13. The range of effective modulus for linear polymers (plastics) is about 100 to 1 over ≈ 12 orders of strain rate, and that for common rubbers is about 1000 to 1 over ≈ 8 orders of strain rate.

The location of the curves on the temperature axis varies with strain rate, and vice versa as shown in Figure 2.13. The temperature–strain rate interdependence, i.e., the amount, a_T, that the curves for E and Δ are translated due to temperature, can be expressed by either of two equations (with varying degrees of accuracy):

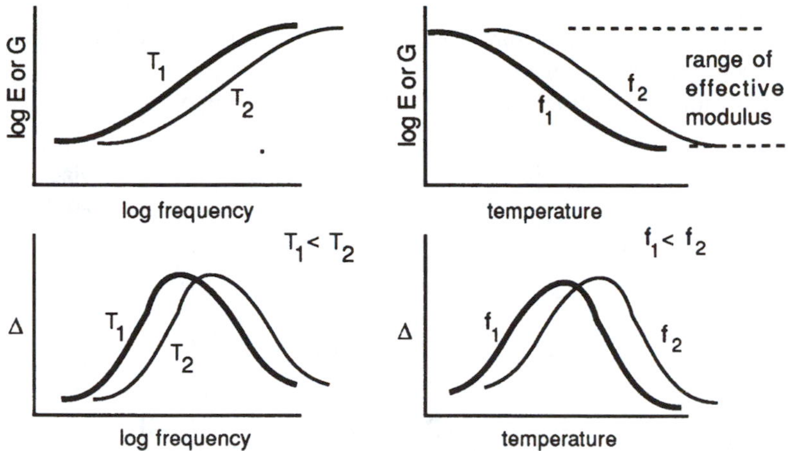

Figure 2.13 Dependence of elastic modulus and damping loss on strain rate and temperature. (Adapted from Ferry, J. D., *Visco-Elastic Properties of Polymers,* John Wiley & Sons, New York, 1961.)

$$\text{Arrhenius: } \log(a_T) = \frac{\Delta H}{R}\left(\frac{1}{T} - \frac{1}{T_o}\right)$$

where ΔH is the (chemical) activation energy of the behavior in question, R is the gas constant, T is the temperature of the test, and T_o is the "characteristic temperature" of the material; or

$$\text{WLF: } \log(a_T) = \frac{-8.86(T - T_s)}{(101.6 + T - T_s)}$$

where $T_s = T_g + 50°C$ and T_g is the glass transition temperature of the polymer.[1]

The glass transition temperature, T_g, is the most widely known "characteristic temperature" of polymers. It is most accurately determined while measuring the coefficient of thermal expansion upon heating and cooling very slowly. The value of the coefficient of thermal expansion is greater above T_g than below. (Polymers do not become transparent at T_g; rather they become brittle like glassy solids, which have short range order. Crystalline solids have long range order; whereas super-cooled liquids have no order, i.e., are totally random.)

An approximate value of T_g may also be marked on curves of damping loss (energy loss during strain cycling) versus temperature. The damping loss peaks are caused by morphologic transitions in the polymer. Most solid (non "rubbery") polymers have 2 or 3 transitions in simple cyclic straining. For example, PVC shows three peaks over a range of temperature. The large (or α) peak is the most significant, and the glass transition is shown in Figure 2.14. This transition is thought to be the point at which the free volume within the polymer becomes greater than 2.5% where the molecular backbone has room to move freely. The secondary (or β) peak is thought to be due to transitions in the side chains. These take place at lower temperature and therefore at smaller free volume since the side chains require less free volume to move. The third (or γ) peak is thought to be due to adjacent hydrogen bonds switching positions upon straining.

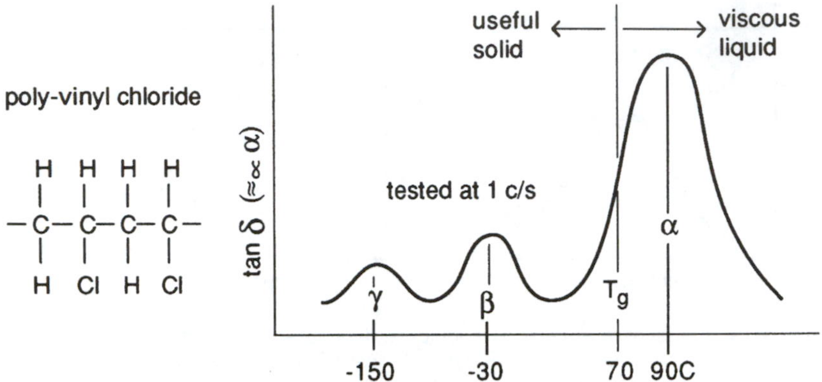

Figure 2.14 Damping loss curve for polyvinyl chloride.

The glass–rubber transition is significant in separating rubbers from plastics: that for rubber is below "room" temperature, e.g., –40°C for the tire rubber, and that for plastics is often above. The glass transition temperature for polymers roughly correlates with the melting point of the crystalline phase of the polymer.

The laboratory data for rubber have their counterpart in practice. For a rubber sphere the coefficient of restitution was found to vary with temperature, as shown in Figure 2.15. The sphere is a golf ball.[2]

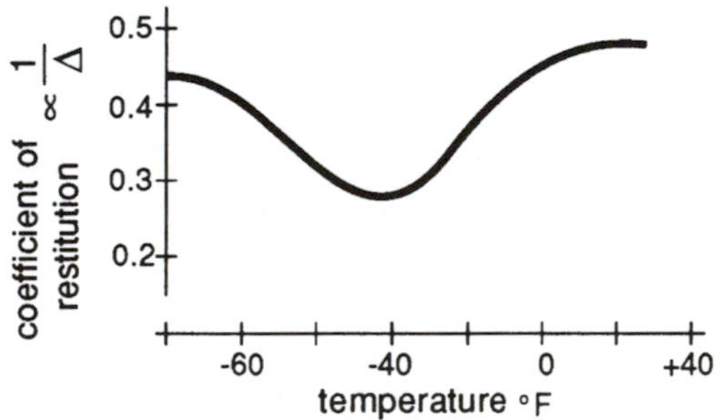

Figure 2.15 Bounce properties of a golf ball.

An example of visco-elastic transforms of friction data by the WLF equation can be shown with friction data from Grosch (see Chapter 6 on polymer friction). Data for the friction of rubber over a range of sliding speed are very similar in shape to the curve of Δ versus strain rate shown in Figure 2.13. The data for μ versus sliding speed for acrylonitrilebutadiene at 20°C, 30°C, 40°C, and 50°C are shown in Figure 2.16, and the shift distance for each, to shift them to T_s is calculated.

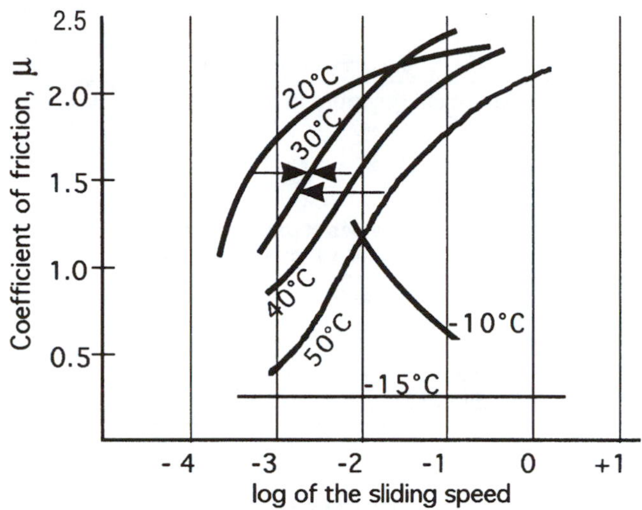

Figure 2.16 Example of WLF shift of data.

For this rubber, $T_g = -21°C$, thus $T_s = +29$, and $\log(a_T) = \dfrac{-8.86(T - T_s)}{101.6 + T - T_s}$

To transform the 50°C data, $\log(a_T) = \dfrac{-8.86(50-29)}{101.6+(50-29)} = \dfrac{-8.86 \times 21}{101.6+21} = -1.51$

i.e., the 50°C curve must be shifted by 1.51 order of 10, or by a factor of 13.2 to the left (negative log a_T) as shown. The 40°C curve moves left, i.e., $10^{0.87}$, the 30°C curve remains virtually where it is, and the 20°C curve moves to the right an amount corresponding to $10^{0.86}$.

When all curves are so shifted then a "master curve" has been constructed which would have been the data taken at 29°C, over, perhaps 10 orders of 10 in sliding speed range.

(See Problem Set question 2 f.)

DAMPING LOSS, ANELASTICITY, AND IRREVERSIBILITY

Most materials are nonlinearly elastic and irreversible to some extent in their stress–strain behavior, though not to the same extent as soft polymers. In the polymers this behavior is attributed to dashpot-like behavior. In metals the reason is related to the motion of dislocations even at very low strains, i.e., some dislocations fail to return to their original positions when external loading is removed. Thus there is some energy lost with each cycle of straining. These losses are variously described (by the various disciplines) as hysteresis losses, damping losses, cyclic energy loss, anelasticity, etc. Some typical numbers for materials are given in Table 2.2 in terms of

$$\Delta = \frac{\text{energy loss per cycle}}{\text{strain energy input in applying the load}}$$

Table 2.2 Values of Damping Loss, Δ for Various Materials

steel (most metals)	≈0.02 (2%)
cast iron	≈0.08
wood	≈0.03–0.08
concrete	≈0.09
tire rubber	≈0.20

HARDNESS

The hardness of materials is most often defined as the resistance to penetration of a material by an indenter. Hardness indenters should be at least three times *harder* than the surfaces being indented in order to retain the shape of the indenter. Indenters for the harder materials are made of diamonds of various configurations, such as cones, pyramids, and other sharp shapes. Indenters for softer materials are often hardened steel spheres. Loads are applied to the indenters such that there is considerable plastic strain in ductile metals and significant amounts of plastic strain in ceramic materials. Hardness numbers are somewhat convertible to the strength of some materials, for example, the Bhn_{3000} (Brinell hardness number using a 3000 Kg load) multiplied by 500 provides a fair estimate of the tensile strength of steel in psi (or use Bhn × 3.45 ≈TS, in MPa).

The size of indenter and load applied to an indenter are adjusted to achieve a compromise between measuring properties in small homogeneous regions (e.g., single grains which are in the size range from 0.5 to 25 μm diameter) or average properties over large and heterogeneous regions. The Brinell system produces an indentation that is clearly visible (≈3 – 4 mm); the Rockwell system produces indentations that may require a low power microscope to see; and the indentations in the nano-indentation systems require high magnification microscopy to see. For ceramic materials and metals, most hardness tests are static tests, though tests have also been developed to measure hardness at high strain rates (referred to as dynamic hardness). Table 2.3 is a list of corresponding or equivalent hardness numbers for the most common systems of static hardness measurement.

Polymers and other visco-elastic materials require separate consideration because they do not have "static" mechanical properties. Hardness testing of these materials is done with a spring-loaded indenter (the Shore systems, for example). An integral dial indicator provides a measure of the depth of penetration of the

Table 2.3 Approximate Comparison of Hardness Values as Measured by the Most Widely Used Systems (applicable to steel mostly)

	Brinell	Rockwell			Vickers
	3000 kg, 10mm ball	b 1/16" ball 100 kg f	c cone 150 kg f	e 1/8" ball 60 kg f	diamond pyramid 1–120 g
		10		62	↑
		20		68	
		30		75	
		40		81	
		50		87	same as
	100	60		93	Brinell
	125	71		100	
	150	81			
	175	88	7		
	200	94	15		
	225	97	20		
	250	102	24		↓
	275	104	28		276
	300		31		304
	325		34		331
	350		36		363
	375		38		390
requires carbide ball	400		41		420
	450		46		480
	500		51		540
	550		55		630
	600		58		765
	650		62		810
	675		63		850
	700		65		940
	750		68		1025

Comparisons will vary according to the work hardening properties of materials being tested. Note that each system offers several combinations of indenter shapes and applied loads.

indenter in the form of a hardness number. This value changes with time so that it is necessary to report the time after first contact at which a hardness reading is taken. Typical times are 10 seconds, 30 seconds, etc., and the time should be reported with the hardness number. Automobile tire rubbers have hardness of about 68 Shore D (10 s).

Notice the stress states applied in a hardness test. With the sphere the substrate is mostly in compression, but the surface layer of the flat test specimen is stretched and has tension in it. Thus one sees ring cracks around circular indentations in brittle material. The substrate of that brittle material, however, usually plastically deforms, often more than would be expected in brittle materials. In the case of the prismatic shape indenters, the faces of the indenters push materials apart as the indenter penetrates. Brittle material will crack at the apex of the polygonal indentation. This crack length is taken by some to indicate the brittleness, i.e., the fracture toughness, or stress intensity factor, K_c. (See the section on *Fracture Toughness* later in this chapter.)

Hardness of minerals is measured in terms of relative scratch resistance rather than resistance to indentation. The Mohs Scale is the most prominent scratch *hardness* scale, and the hardnesses for several minerals are listed in Table 2.4.

(See Problem Set question 2 g.)

RESIDUAL STRESS

Many materials contain stresses in them even though no external load is applied. Strictly, these stresses are not material properties, but they may influence apparent properties. Bars of heat-treated steel often contain tensile residual stresses just under the surface and compressive residual stress in the core. When such a bar is placed in a tensile tester, the applied tensile stresses add to the tensile residual stresses, causing fracture at a lower load than may be expected.

Compressive residual stresses are formed in a surface that has been shot peened, rolled, or burnished to shallow depths or milled off with a dull cutter. Tensile residual stresses are formed in a surface that has been heated above the recrystallization temperature and then cooled (while the substrate remains unheated). Residual stresses imposed by any means will cause distortion of the entire part and have a significant effect on the fatigue life of solids.

(See Problem Set question 2 h.)

FATIGUE

Most material will fracture when a small load is applied repeatedly. Generally, stresses less than the yield point of the material are sufficient to cause fatigue fracture, but it may require between 10^5 and 10^7 cycles of strain to do so. Gear teeth, rolling element bearings, screws in artificial hip joints, and many other mechanical components fail by elastic fatigue. If the applied cycling stress exceeds the yield point, as few as 10 cycles will cause fracture, as when a wire coat hanger is bent back and forth a few times. More cycles are required if the strains per cycle are small. Failure due to cycling at stresses and strains above the yield point is often referred to as low-cycle fatigue or plastic fatigue.

There is actually no sharp discontinuity between elastic behavior and plastic behavior of ductile materials (dislocations move in both regimes) though in high cycle or elastic fatigue, crack nucleation occurs late in the life of the part, whereas in low-cycle fatigue, cracks initiate quickly and propagation occupies a large fraction of part life. Wöhler (in reference number 3) showed that the entire behavior of metal in fatigue could be drawn as a single curve, from a low stress at which fatigue failure will never occur, to the stress at which a metal will fail in a quarter cycle fatigue test, i.e., in a tensile test. A Wöhler curve for constant strain amplitude cycling is shown in Figure 2.17 (few results are available for the more difficult constant stress amplitude cycling).

There are several relationships between fatigue life and strain amplitude available in the literature. A convenient relationship is due to Manson (in reference number 3) who suggested putting both high-cycle fatigue and low-cycle fatigue into one equation:

Table 2.4 Mohs Scale of Scratch Hardness

	O	E	(Equiv. Knoop)	Reference Minerals	
talc	**1**			hydrous mag. silicate	$Mg_3Si_4O_{10}(OH)_2$
carbon, soft grade	1.5				
boron nitride	≈2			(hexagonal form)	
fingernail	>2				
gypsum	**2**		32	hydrated calcium sulfate	$CaSO_4 \cdot 2H_2O$
aluminum	≈2.5				
ivory	2.5				
calcite	**3**		135	calcium carbonate	$CaCO_3$
calcium fluoride	4		163		
fluorite	**4**			calcium fluoride	CaF_2
zinc oxide	4.5				
apatite	**5**		430	calcium fluorophosphate	$Ca_3P_2O_8$-CaF_2
germanium	≈5				
glass, window	>5				
iron oxide	5.5 to 6.5			rouge	
magnesium oxide	≈6			periclase	
orthoclase	**6**		560	potassium aluminum silicate	$KAlSi_3O_8$
rutile	>6			titanium dioxide	TiO_2
tin oxide	6 to 7			putty powder	
ferrites	7 to 8				
quartz	**7**	8	820	silicon dioxide,	SiO_2
silicon	≈7				
steel, hardened	≈7				
chromium	7.5				
nickel, electroless	8				
sodium chloride	>8			NaCl	
topaz	**8**	9	1340	aluminum fluorosilicate	$Al_2F_2SiO_4$
garnet		10			
fused zirconia		11			
aluminum nitride	≈9				
alumina	**9**	12		alpha, corundum,	Al_2O_3
ruby/sapphire	9		1800		
silicon carbide	>9	13		alpha, carborundum	
silicon nitride	≈9				
boron carbide		14	4700		
boron nitride (cubic)		≈14.5			
diamond	**10**	15	7000	carbon	

O signifies original Mohs scale with basic values underlined and bold; E signifies the newer extended range Mohs scale. The original Mohs number ≈0.1 R_c in midrange, and the new Mohs numbers ≈0.7(Vickers hardness number)[1,3]

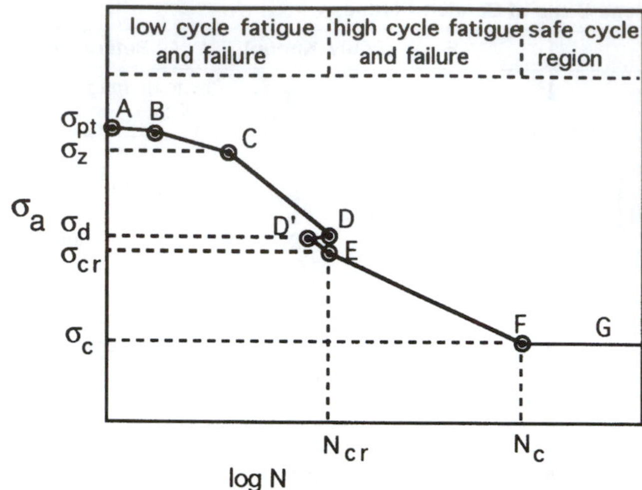

Figure 2.17 Curve by Wöhler showing the connection between all modes of fatigue behavior.

σ_{pt} = true fracture stress in tension
σ_z = stress at first signs of fatigue failure at the surface
σ_d = stress at the occurrence of discontinuity in the Wöhler curve
σ_{cr} = critical stress between low-cycle fatigue and high-cycle fatigue
σ_c = fatigue limit
A-D = region of low-cycle fatigue
A-B and B-C = the failure is of quasistatic character
B-C = region of ratchetting in low-cycle fatigue
C-D = in addition to the quasistatic failure, characteristic areas of fatigue failure can be observed on the fracture surface
D-D′ = transition region between the two types of fatigue failure
D′-E-F = region of high-cycle fatigue
F-G = region of safe cyclic loading

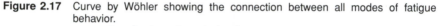

$$\frac{\Delta\varepsilon_t}{2} = \frac{\Delta\varepsilon_p}{2} + \frac{\Delta\varepsilon_e}{2} = \varepsilon_f'(2N_f)^c + \frac{\sigma_f'}{E}(2N_f)^b$$

where N_f = number of cycles to failure, the conditions of the test are:
$\Delta\varepsilon_p$ = plastic strain amplitude
$\Delta\varepsilon_e$ = elastic strain amplitude
$\Delta\varepsilon_t$ = total strain amplitude
and the four fatigue properties of the material are:
b = fatigue strength exponent (negative)
c = fatigue ductility exponent (negative)
σ'_f = fatigue strength coefficient
ε'_f = fatigue ductility coefficient

This equation may be plotted as shown in Figure 2.18, with the elastic and plastic components shown as separate curves. In this figure, $2N_t$ is the transition fatigue life in reversals (2 reversals constitute 1 cycle), which is defined as N_f

for the condition where elastic and plastic components of the total strain are equal. Conveniently, in the plastic range the low-cycle fatigue properties may be designated with only two variables, ε'_f and c (for a given $\Delta\varepsilon_p$).

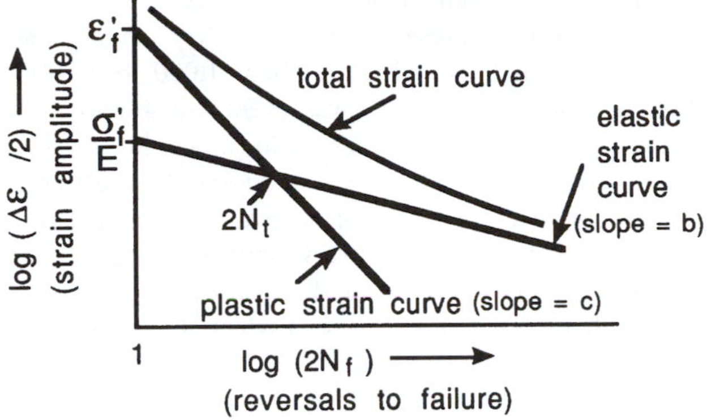

Figure 2.18 Curves for low-cycle fatigue, high-cycle fatigue, and combined mechanisms, in constant strain amplitude testing.

The measuring of low-cycle fatigue properties is tedious and requires specialized equipment. Several methods are available for approximating values of ε'_f and c from tensile and hardness measurements. Some authors set $\varepsilon'_f = \varepsilon_f$ and:

$$c = \frac{\log\left(\dfrac{\sigma'_f}{E\varepsilon'_f}\right) + b \cdot \log(2N_t)}{\log(2N_t)}$$

FRACTURE TOUGHNESS

One great mystery is why "ductile" materials sometimes fracture in a "brittle" manner and why one must use a property of materials known as K_c to design against brittle fracture. Part of the answer is seen in the observation that large structures are more likely to fail in a brittle manner than are small structures. Many materials do have the property, however, of being much less ductile (or more brittle, to refer to the absence of a generally useful attribute) at low temperatures than at higher temperatures. Furthermore, when high strain rates are imposed on materials as by impact loading, many materials fracture in a brittle manner. It was to examine the latter property that impact tests were developed, such as the Charpy and Izod tests, for example. These tests measure a quantity somewhat related to area under the stress–strain curve (i.e., energy) at the strain rates associated with impact. The major difficulty with these tests is that there is no good way to separate actual fracture energy from the kinetic energy, both of the ejected specimens after impact and in the vibrations in the test machine due to impact.

The mathematics of fracture mechanics appears to have developed from considerations some 60 years ago of the reason why real materials are not as strong as they "should be." Calculations from the forces that exist between atoms at various atom spacing (as represented in Figure 3.1) suggest that the strength of solids should be about E/10, which is about 1,000 to 10,000 higher than practical values. In ductile metals this was eventually found to be due to the influence of dislocation motion. However, dislocations do not move very far in glasses and other ceramic materials. The weakness in these materials was attributed to the existence of cracks, which propagate at low average stress in the body. Fracture mechanics began with these observations and focused on the influence of average stress fields, crack lengths, and crack shapes on crack propagation. Later it was found that the *size* of the body in which the crack(s) is (are) located also has an influence.

Studies in fracture mechanics and fracture toughness (sometimes said to be the same, sometimes not) are often done with a specimen of the shape shown in Figure 2.19. The load P opens the crack by an amount (displacement) δ, making

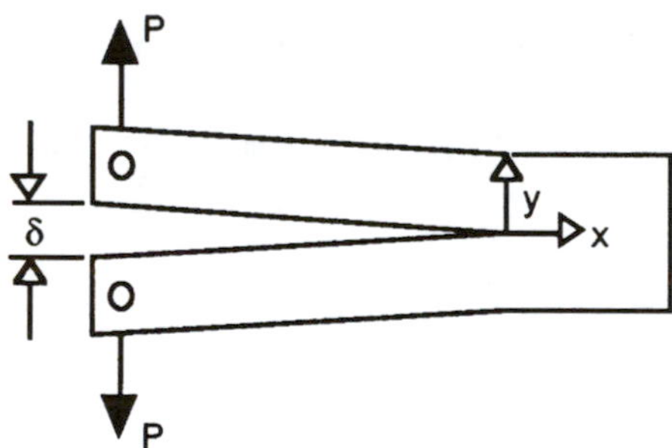

Figure 2.19 The split beam specimen.

the crack propagate in the x direction. As the crack propagates, new surface area is created, which requires an amount of energy equal to twice the area, A, of the crack (two surfaces), multiplied by the surface energy, γ, to form each unit of new area. (When rejoining of the crack walls restores the system to its original state, that energy per unit area is called the surface free energy.) If the crack can be made to propagate quasistatically, $P\delta=2A\gamma$: much mathematics of fracture is based on the principle of this energy balance. The equation,

$$d(\delta/P)/dA = 2R/P^2$$

is used, where the value of R at the start of cracking is called the critical strain energy release rate, i.e., the rate at which A increases.

Another part of fracture mechanics consists of calculating the stresses at the tip of the crack. This is done in three separate modes of cracking, namely, Mode I where P is applied as shown in Figure 2.19; Mode II where P is applied such that the two cracked surfaces slide over each other, left and right; and Mode III where P is applied perpendicular to that shown in Figure 2.19, one "into" the paper and the other "outward." An example of a calculation in Mode I for a plate, 2b wide with a centrally located slit 2L long, in a plate in which the average stress, σ, is applied, has a stress intensity factor, K, of

$$K = \sigma\sqrt{2b\tan^{-1}\left(\frac{\pi L}{2b}\right)}$$

which has the peculiar units of $N/m^{3/2}$ or $lbf/in^{3/2}$. K is not a stress concentration in the sense of a multiplying factor at a crack applied upon the average local stress. Rather, it is a multiplying factor that reflects the influence of the sizes of both the crack and the plate in which the crack is located. Values of K have been calculated for many different geometries of cracks in plates, pipes, and other shapes, and these values may be found in handbooks.

Cracking will occur where K approaches the critical value, K_c, which is a material property. The value of K_c is measured in a small specimen of very specific shape to represent the basic (unmultiplied) part size. In very brittle materials the value of K may be calculated from cracks at the apex of Vickers hardness indentations. The indenter is pyramidal in shape and produces a four-sided indentation as shown in Figure 2.20. Cracks emanate from the four corners to a length of c. The value of K_c is calculated with the equation:

$$K_1 = \xi\left(\frac{E}{H}\right)^{.5} Wc^{\frac{-3}{2}}$$

where W is the applied load and ξ is a material constant, usually about 0.016.

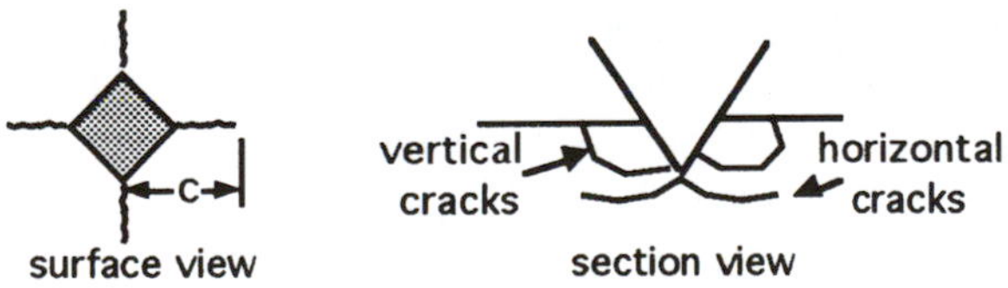

Figure 2.20 Cracks emanating from a hardness indentation.

The consequence of structure size may be seen in Figure 2.21.[4] As the size of the structure increases, K increases. The acceptable level of σ when $K = K_c$ is lower in a large structure than in a small one and becomes lower than σ_y at some point.

The stress required to *initiate* a crack is higher than the stress needed to *propagate* a crack: this difference is very small in glass but large in metal. In ductile materials the crack tip is blunt and surrounded by a zone of plastic flow. Typically, brittle ceramic materials have values of K_c of the order of 0.2 to 10

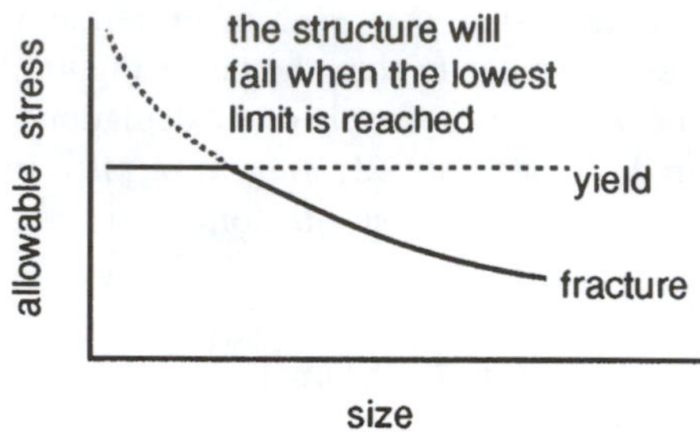

Figure 2.21 A sketch of the influence of structure size on possible types of failure. (Adapted from Felbeck, D.K. and Atkins, A.G., *Strength and Fracture of Engineering Solids,* Prentice Hall, 1984.)

MPa√m, whereas soft steel will have values of the order of 100 to 175. However, as the crack in a large structure of steel begins to propagate faster, the plastic zone diminishes in size (and amount of energy adsorbed diminishes). The crack accelerates, requiring still less energy to propagate, etc.

The calculations above refer to plane strain fields. For plane stress the calculated values will be one third those for plane strain. Correspondingly, K_c will be higher where there is plane stress than where there is plane strain.

APPLICATION TO TRIBOLOGY

All of the above material properties are really responses to stresses applied in rather specific ways. The wearing of material is also a response to applying stresses (including chemical stresses). The mechanical stresses in sliding are very different from those imposed in standard mechanical tests, which is why few of the existing models for material wear adequately explain the physical observations of wear tests. This may be seen by comparing the stress state in a flat plate, under a spherical slider with those in the tests for various material properties. Three locations under a spherical slider are identified by letters a, b, and c in the flat plate as shown in Figure 2.22a. Possible Mohr circles for each point are shown in Figure 2.22b. Note that location b in Figure 2.22a has a stress state similar to that under a hardness indenter.

Circles d and e in Figure 2.22c are for the stress states in a fracture toughness test and in a tensile test, respectively. The fracture toughness test yields values of the critical stress intensity factors, K_c, for fracture, and the tensile test yields Young's Modulus, both of which are found in wear models. Only the approximate axes with the shear and cleavage limits for two different material phases including locations of the Mohr circle for these tests are given. Two observations may be made, namely, that the stresses imposed on material under a slider are very different from those in tensile and fracture toughness tests, and the stress state under a slider varies with time as well. The reader must imagine the mode of failure that will occur as each circle becomes larger due to increased stress. It

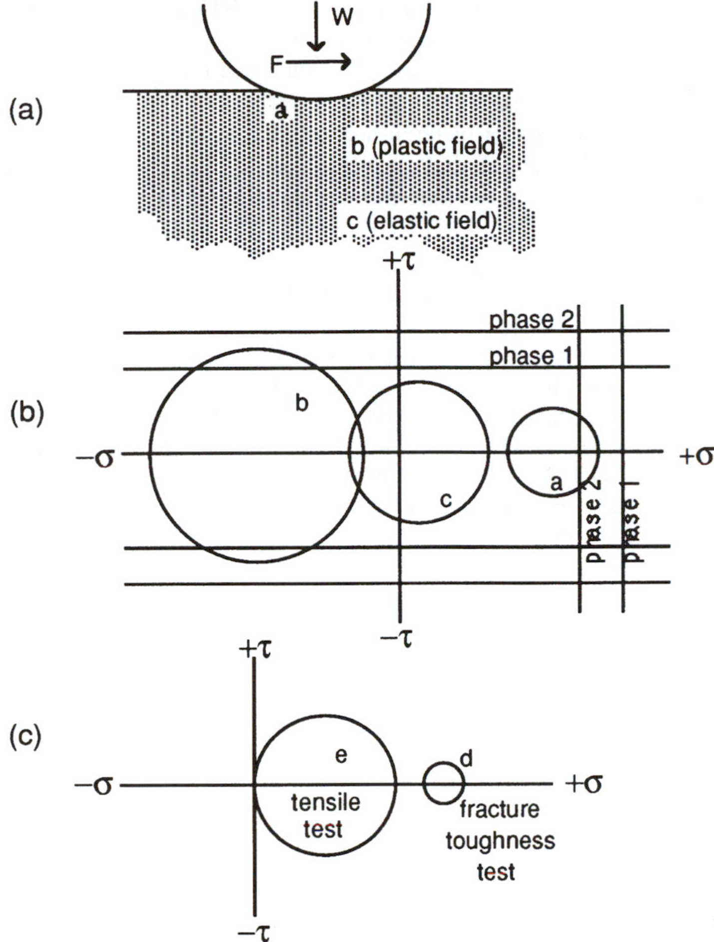

Figure 2.22 Stress state under a spherical slider and five stress states on Mohr circle axes with the shear and cleavage limits for two different material phases included.

may be seen that circle d is not likely to invoke plastic deformation and circle b is not likely to invoke a brittle mode of failure.

It should be noted that the conclusions available from the Mohr circle alone are inadequate to explain the effects of plastic deformation versus brittle failure. The consequence of plastic flow in the strained material is to reconfigure the stress field, either by relieving the progression toward brittle failure, or perhaps by shifting the highest tensile stress field from one phase to another in a two-phase system. Further, plastic flow requires space for dislocations to move (glide). Asperity junctions and grain sizes are of the order of 0.5 to 5μm. If local (contact) stress fields are not oriented for easy and lengthy dislocation glide, or for easy cross slip, that local material will fracture at a small strain, but may resist fracture as if it had a strength 10 to 100 times that of the macroscopic yield strength.

Figure 2.22b also shows the shear and cleavage limits of two different materials that may exist in a two-phase material. Frequently, one phase is "ductile," in which the shear limit is less than half the cleavage limit, and the other phase is "brittle," showing the opposite behavior.

An important property of material not included in Figure 2.22 is the fatigue limit of materials. Perhaps fatigue properties could be shown as a progressive reduction in one or both of the failure limits with cycles of strain.

REFERENCES

1. Ferry, J.D., WLF = Williams, Landell and Ferry, in *Visco Elastic Properties of Polymers*, John Wiley & Sons, New York, 1961.
2. U.S. Bureau of Standards, *J. of Research*, 34, 19, 1945.
3. Pushkar, A. and Golovin, S.A., *Fatigue in Materials: Cumulative Damage Processes*, Material Science Monograph 24, Elsevier, 1985.
4. Felbeck, D.K. and Atkins, A.G., *Strength and Fracture of Engineering Solids*, Prentice Hall, 1984.

CHAPTER 3

Adhesion and Cohesion Properties of Solids: Adsorption to Solids

Perhaps the most misleading comment in the mechanics of tribology relating to the instant of contact is, "and there is adhesion," apparently implying bonding of uniformly high strength over the entire contact area. It is not that simple in the vast majority of contacting events. Ever-present but ill-defined adsorbed gases and contaminants, as well as the directional properties of atomic bonds, limit attachment strength to low values.

INTRODUCTION

Aggregates, clumps, or groups of atoms are all generally attracted toward each other just as the planets and stars are. Bonding between atoms may be described in terms of their electron structure. In the current shell theory of electrons it would appear that the number of electrons with negative charge would balance the positive charge on the nucleus and there would be no net electrostatic attraction between atoms. However, within clusters of atoms the valence electrons (those in the outer shells) take on two different duties. In the covalent bond, for example, a pair of electrons orbit around two adjacent atoms and constitute the "s" bond. The remaining electrons in nonconductors, and all valence electrons in metal, become "delocalized," setting up standing waves among a wide group of nuclei, forming the π bond. The average energy state of these delocalized electrons is lower than the energy state of valence electrons in single atoms, and this is the energy of bonding between atoms. These energy states can be detected most readily by spectroscopic measurements.

ATOMIC (COHESIVE) BONDING SYSTEMS

There are four atomic bonding systems in nature: the metallic bond, the ionic bond, the covalent bond, and the van der Waals bond systems. These are often referred to as cohesive bonding systems.

The Metallic (or Electronic) Bond: Those elements that readily conduct heat and electricity are referred to as metals. The valence electrons of metallic elements are not bound to specific nuclei as they are in ceramic and polymeric materials. Coincidentally, the variation in bonding energy, as a single atom moves along a "flat" array of other atoms, is small. The atoms are therefore not highly constrained to specific locations or bond angles relative to other atoms.

The Covalent Bond: When two or more atoms (ions of the same charge) share a pair of electrons such that they constitute a stable octet, they are referred to as covalently bonded atoms. For example, a hydrogen atom can bond to one other hydrogen or fluorine or chlorine (etc.) atom because all of these have the same number of valence electrons (+ or –). Some single atoms will have enough electrons to share with two or more other atoms and form a group of strongly attached atoms. Oxygen and sulfur have two covalent bonds, nitrogen has three, carbon and silicon may have four. To dislodge covalently bonded atoms from their normal sites requires considerable energy, almost enough to separate (evaporate) the atoms completely. The bond angles are very specific in covalent solids.

The carbon–carbon bond, as one covalent material, may produce a three-dimensional array. In this array the bonds are very specific as to angle and length. This is why diamond is so hard and brittle.

When a single atom is brought down to a plane containing covalently bonded atoms, the single atom may receive either very little attention, or considerable attention depending on the exact site upon which it lands. Two planes of three-dimensional covalently bonded atoms will adhere very strongly if the atoms in the two surfaces happen to line up perfectly, but if each surface is a different lattice plane or if identical lattice planes are rotated slightly, the adhesion will be considerably reduced, to as low as 3% of the maximum.

The Ionic Bond: Some materials are held together by electrostatic attraction between positive and negative ions. Where the valence of the positive and negative ions is the same, there will be equal numbers of these bonded ions. Where, for example, the positive ion has a larger charge than do surrounding negative ions, several negative ions will surround the positive ion, consistent with available space between the ions. (Recall that the positive ion will usually be smaller than the negative ion.) Actually, the ion pairs or clusters do not become isolated units. Rather, all valence electrons are π electrons, that is, the valence electrons vibrate in synchronization with those in adjacent electrons, binding the atoms together.

Ionic bonds are very strong. They can accommodate only a little more linear and angular displacement than can the covalent bonds. Again, two surfaces of ionic materials may adhere with high strength, or a lower strength depending on the lattice alignment.

[Crystal structure is determined by a combination of the number of ions needed for group neutrality and optimum packing. Many atomic combinations cannot be accommodated to satisfy covalent or ionic bonding structures. For example, diamond is 100% covalent, SiC is 90% covalent and 10% ionic, Si_3N_4 is 75% and SiO_2 is 50% covalent.]

Molecules: Molecules are groups of atoms usually described by giving examples. Generally, crystalline and lamellar solids (groups of atoms) are not referred to as molecular. Several different molecules may be made up of the same atoms, such as nitrogen, oxygen, or chlorine gases. Three types of hydrocarbon molecules are shown in the sketch below:

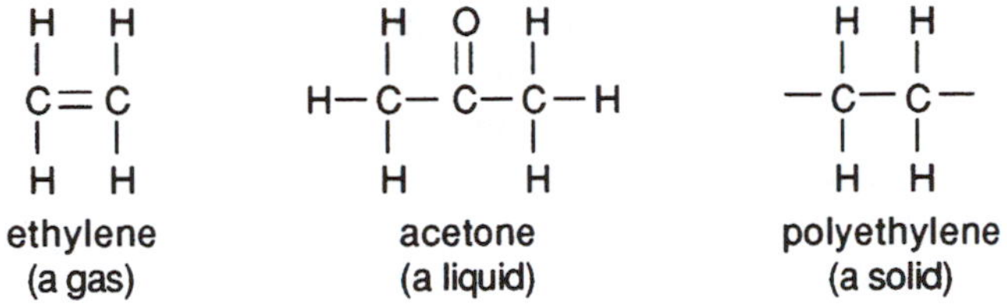

```
 H   H           H  O  H            H  H
 |   |           |  ||  |            |  |
 C = C       H - C - C - C - H    - C - C -
 |   |           |       |           |  |
 H   H           H       H           H  H
ethylene         acetone          polyethylene
(a gas)         (a liquid)         (a solid)
```

These three molecules are based on the carbon atom. Carbon has four bonds which are represented by lines, the single line for the single (strength) bond and the double lines for the double (strength) bond. Hydrogen has one bond and oxygen has two.

Within the molecule, the atoms are firmly bonded together and are arranged with specific but compliant bond angles. Actually, the molecules are not two dimensional, but rather each CH_2 unit is rotated a certain amount relative to adjacent ones around the carbon bond. These molecules are not completely independent units, but rather are bonded together by the weak forces of all nearby resonating electrons. Note that the center of positive charge in the acetone molecule coincides with the middle C atom, whereas the oxygen ion carries a negative charge. This separation of charge centers makes the acetone molecule a polar molecule. The other two molecules are nonpolar.

The van der Waals Bonds: Attractive forces of atoms extend a distance of 3 or 4 times the radius of an atom, though the forces at this distance are weak. When atoms are assembled as molecules these forces are enhanced in proportion to the size of molecules, and enhanced further by any polarity that exists in some molecules. In large molecular structures such as the polymers, these forces bind the molecules together and constitute a major part of the strength of the polymer material. The strength is much less than that of the ionic, covalent, and metallic bonds however.

ADHESION

Bonding Between Dissimilar Materials Within the Same Bond Classification: The discussions on atomic bonding often focus on simple systems. In engineering practice, parts sliding against each other are often dissimilar. A brass sliding on steel, with no adsorbed layers present, might be expected to bond according to the rules of the metallic bond system, and similarly with the covalent, ionic, and van der Waals systems. All cleaned metals that have been contacted together in vacuum have bonded together with very high strength. It is possible that solubility of one metal in the other may enhance adhesion and thereby influence friction (and wear) but not significantly at temperatures below two thirds of the MP in absolute units.

Adhesion experiments with ceramic materials have not yielded high bond strength, probably because of the difficulty in matching lattices as perfectly as required. However, when two different ceramic materials are rubbed together, there is an increased probability that some fortuitous and adequate alignment of lattices occurs to form strong bonds. Debris is also formed and these particles also bond to one or another of the sliding surfaces. Layers of debris sometimes form such compact films as to reduce the wear rate.

Disparate Bonds: The term "disparate" bonds is an unofficial classification, used here to refer to the bonding that takes place between a covalent system and an ionic system, or between an ionic and metallic system, etc. For example, the bond strength between a layer of Al_2O_3 "grown on" aluminum is very high though Al_2O_3 is an ionic ceramic material and aluminum is a metal. Again, when polyethylene is rubbed against clean glass or metal, a film of the polymer is left behind, indicating that the (adhesive) bond between the polymer film and glass or metal is about as strong as the (cohesive) bonds within the polymer. In general some disparate systems might be expected to bond well because the surfaces of all materials have different structures and energy states than do the interiors. Where there is reasonable lattice matching there could then be high bond strength. This is the subject of current research in materials science, and few guidelines are yet available.

ATOMIC ARRANGEMENTS: LATTICE SYSTEMS

The energy of bonding, and therefore the bonding forces, vary with distance between pairs of atoms, which can be schematically represented as shown in Figure 3.1. The net force, or energy, is usually described as the sum of two forces, an attraction force and a repulsion force. The force of attraction is related to the inverse of the square of the distance of the separation of the charges. The force of repulsion arises from attempting to place too many electrons in closer than "normal" proximity.

Atoms in a large three-dimensional array cannot be arranged with zero force between them. Rather, the nearest neighbors are too close and the next nearest are farther apart than the spacing which produces zero force. The result is that atoms will stack in 14 very specific three-dimensional arrays, according to the "size" of atoms and the forces between atoms at specific spacing. Most metals are arranged in either the body-centered cubic, the face-centered cubic, or the hexagonal close-packed array. These three arrays are shown in Figures 3.2 and 3.3.

Table 3.1 lists the common metals according to their lattice arrangements. These and a few other arrays are also found in ionic and covalent materials. The

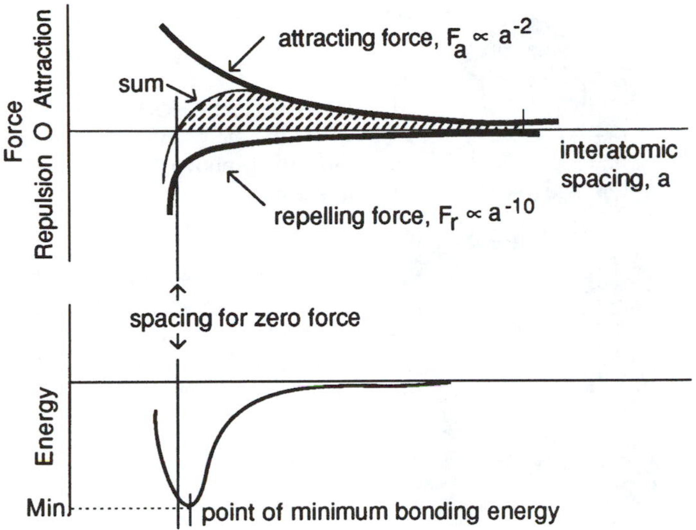

Figure 3.1 Schematic representation of the forces and energy between atoms.

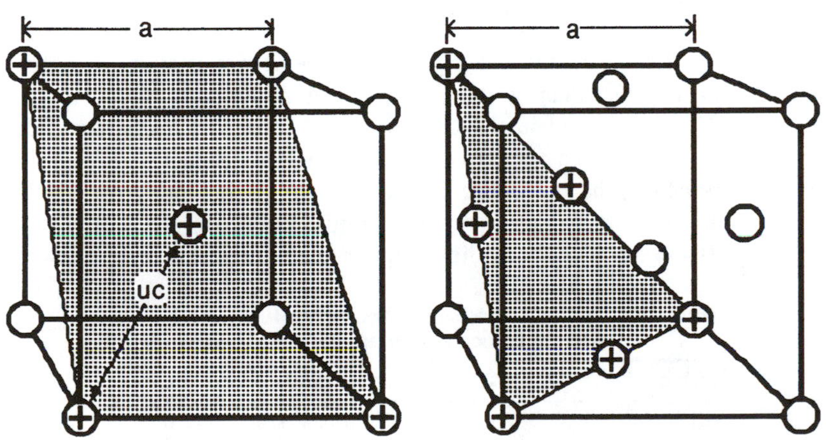

body-centered cubic (BCC) array of atoms with one atom at the corners of the cube and one in the center of the cube on the cross - hatched face diagonal plane

face-centered cubic (FCC) array of atoms with one atom at the corners of the cube and one in the center of each of six faces. The cross-hatched plane is one of eight octahedral planes.

Figure 3.2 Atomic arrangement in the body-centered and face-centered cubic lattice arrays. The cubic array is one of several ways to designate the position of atoms. For some purposes the unit cell (uc) is identified. The uc for the FCC array is composed of the atom in one corner plus the atoms in the center of adjacent faces. For still other purposes a set of the cross-hatched planes is used to indicate the direction in which crystals will shear.

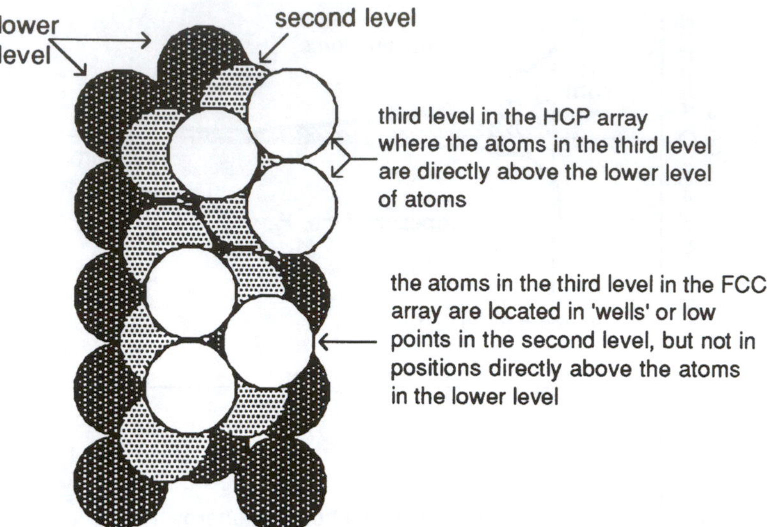
lower level
second level
third level in the HCP array where the atoms in the third level are directly above the lower level of atoms
the atoms in the third level in the FCC array are located in 'wells' or low points in the second level, but not in positions directly above the atoms in the lower level

Figure 3.3 Atomic stacking in the face-centered cubic and hexagonal close-packed lattice array. The face-centered cubic (FCC) and the hexagonal close-packed (HCP) arrays differ from each other in the "stacking" of the octahedral or body-diagonal planes. Atoms on the octahedral planes are shown for two arrays.

size of atoms is defined by the spacing between the center of atoms in a three-dimensional array rather than by the size of the outermost electron shells. Iron atoms at 20°C are arranged in the body-centered cubic (BCC) lattice with a corner-to-corner distance, a = 0.286 nm. The smallest distance between atom centers occurs across the body diagonal (diagonally across the cross hatched plane in Figure 3.2) where there are four atomic radii covering a distance of $\sqrt{3} \times 0.286 = 0.495$ nm. Thus the radius of the iron atom is 0.433 a, or 0.124 nm.

Table 3.1 List of Some Metals According to Their Atomic Lattice Arrangement

Trigonal	FCC	BCC	HCP
Bi	Al	Fe (below 910°C)	Cd
Sb	Cu	Cr	Zn
	Ni	Nb	Mg
	Co	V	Ti
	Fe (above 910°C)	Ta	Zr
		Mo	
		W	

The lattice structure of ceramic materials is much more complicated because of the great difference in size between the anions and cations.

The size of atoms changes either when combined with atoms other than their own type, or when their neighbors are removed. The iron atom when combined with oxygen as FeO has a radius of 0.074 nm and when combined with oxygen as Fe_2O_3 has a radius of 0.064 nm. These are referred to as ion radii. The iron ion has a positive charge and is smaller than the atom. A negative ion is larger than the same atom. Thus the oxygen ion in oxide is larger than the oxygen atom and, further, in oxide the oxygen ion has a larger radius than does iron, ≈ 0.140 nm.

The iron atom in the body-centered cubic form has eight neighbors. Just above 910°C, pure iron is arranged in a face-centered cubic (FCC) lattice array, with a corner-to-corner distance of a = 0.363 nm. The atoms across the face diagonal are spaced most closely, producing an atom radius of 0.128 nm. The FCC atoms have 12 near neighbors.

(See Problem Set question 3 a.)

DISLOCATIONS, PLASTIC FLOW, AND CLEAVAGE

Crystalline structures in commercial materials usually contain many defects. Some of the defects are missing atoms, or perhaps excess atoms, singly or in local groups. One type of defect is the dislocation in the crystalline order. The edge dislocation may be shown as an extra plane as shown in Figure 3.4. Orderly crystal structure exists above, below, and to the sides of the dislocation. When a shear stress, τ, is imposed, large groups of atoms need not be translated in order to achieve movement to the next equilibrium position. Rather atom "a" moves into alignment with atom "b," and atom "c" becomes the unattached end of a plane. This process continues and the dislocation (extra plane) continues to move to the left. Much less shear stress is required for stepwise, single atom displacement than if all atoms were to be displaced at once, actually by about a factor of 1000. The presence of movable dislocations in metal makes them ductile. When the motion of dislocations is impeded by alloy atoms or by entanglement (e.g., due to previous cold work) with other dislocations, a greater shear stress is required to move them: the metal is harder and less ductile. When there are no dislocations, as in a perfect crystal, or where dislocations are immobile as in a ceramic material, the material is brittle.

In Figure 3.4 a stress, σ, is applied in such a way that it cannot induce a shear stress to activate the dislocation. A sufficiently high value of stress will simply separate planes of atoms. If this separation occurs along large areas of the simple crystallographic planes it is called cleavage. Actually separation can occur along any average direction, still occurring along atomic planes.

If the stress, σ, were applied at a 45° angle relative to τ, there would be a normal force applied along atomic planes to cause cleavage and a shear force to move dislocations. Cleavage strength and shear strength are seen as two independent properties of materials.

(See Problem Set question 3 b.)

ADHESION ENERGY

Surface atoms of all arrays have fewer neighbors than do those submerged in a solid, depending on the lattice plane that is parallel to the surface plane. If

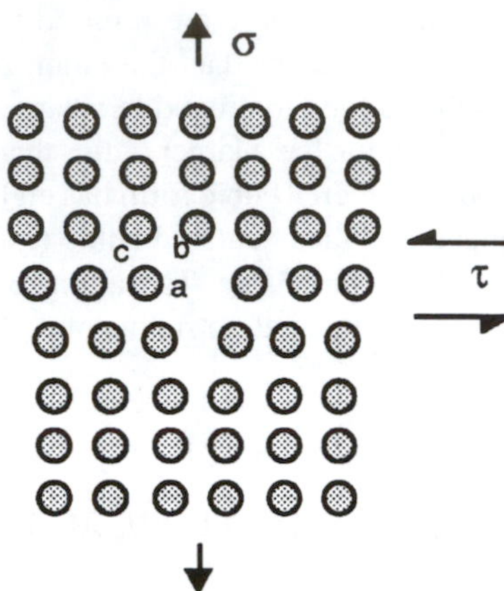

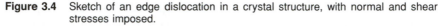

Figure 3.4 Sketch of an edge dislocation in a crystal structure, with normal and shear stresses imposed.

the surface plane is parallel to the "cube face" in the face-centered cubic array, a surface atom has only eight near neighbors, having been deprived of four of them. Surface atoms exist in a higher state of energy and are "smaller" than substrate atoms. Out-of-plane adjustments are made to retain a structure that is somewhat compatible with the face-centered cubic substrate.

The higher state of energy of surface (and near surface) atoms is achieved by adding energy from outside to separate planes of atoms. That energy can be recovered by replacing the separated atoms, which is directly analogous to bringing magnets (of opposite polarity) into and out of contact. This process may not be totally irreversible if some irreversible deformation and defect generation has taken place.

In the perfectly reversible process, the energy exchange is referred to as the *surface free energy.* Where there is some irreversibility in the process, the (new) surface has increased its *surface energy,* some of which may be recovered by replacing the separated body, but not all. The recovery of any amount of energy by replacing the separated body is the basis for adhesion.

ADSORPTION AND OXIDATION

The process by which atoms or molecules of a gas or liquid become attached to a solid surface is called adsorption. The surface of a solid has some unsatisfied bonds which can be satisfied by bringing *any* atom into the area of influence of the unsatisfied bond. Adsorption is always accompanied by a decrease in surface energy.

There are two classes of adsorption, namely, physical and chemical. Physical adsorption, involving van der Waals forces, is found to involve energies of the order of magnitude of that for the liquefaction of a gas, i.e., $Q < 0.2$ KJ/mol

(1 J/mol = 4.19 Cal/mol) in the equation, (reaction rate) R $\propto e^{-Q/RT}$ (the gas constant R is two thirds the total energy of translation of a gas at 1°K) and is easily reversible (varies with temperature and boiling point of gas). Chemisorption involves an energy of activation of the order of chemical reactions, i.e., 2.5 to 25 KJ/mol, because it involves change in chemical structure. It is irreversible, or reversible with great difficulty. Actually, chemisorption involves two steps — physical adsorption followed by the combining of the adsorbate with substrate atoms to form a new compound.

There are several theories and a number of isotherms indicating whether or not, and how vigorously, various adsorption processes may take place. For this purpose one can also use handbook values of the heats of formation compounds formed from gases, as shown in Table 3.2. For example, oxygen settling on copper liberates ΔH = 8.33 KJ/mol when a mole of (cupric) CuO is formed, and 9.52 when (cuprous) Cu_2O is formed. Copper nitride is not listed, so nitrogen very likely forms only a physically adsorbed layer.

The existence of attached gas and nonmetallic or intermetallic layers on solid surfaces is beyond dispute: we do not yet have these layers well enough characterized to estimate their influence in friction, particularly in dry friction.

ADSORBED GAS FILMS

A solid surface, once formed and not yet exposed to other atoms, is very reactive. Impinging atoms or molecules will readily attach or adsorb. In a normal atmosphere of gases including water vapor, layers of gas settle down on the surface and become about 70% as dense as the liquefied or condensed form of the gas. (The oxygen in the layer later forms oxide on metals.) This complex layer shields or masks potentially high adhesion forces between contacting solids and significantly influences friction and wear. The most mysterious characteristic of the literature on the mechanics of friction and wear is the near total absence in consideration of adsorbed films, in the face of overwhelming evidence of the ubiquitous nature of adsorbed films. Perhaps the problem is that the films are invisible. The films do form very quickly. Following is a calculation to show how quickly a single layer forms.

Begin with the assumption of Langmuir that only those molecules that strike a portion of the surface not already covered will remain attached; all others will reevaporate (i.e., sticking factor of 1). The rate of condensation at any time is then $\rho = \rho_o(1 - \theta)$ where ρ_o is the original rate of condensation and $\theta = N/N_o$ where N is the number of molecules per unit area previously settled on the surface and N_o is the maximum number that can be contained per unit area as a single layer. Now ρ is the rate of change in the number of condensed atoms per unit area; i.e.,

$$\rho = dN/dt, \text{ which } = N_o d\theta/dt.$$

Substitution yields: $\rho_o(1 - \theta) = N_o d\theta/dt$

for which the solution is $\mathcal{L}n\,(1 - \theta) = -\,\rho_o\, t/N_o.$

Table 3.2 Some Properties of Common Elements

	Element	Young's Modulus GPa	Density ..(g/cc)..	MP°C	BP°C	Thermal conduct. ($^{J}/_{s}$ cm °K)	Oxide	ΔH, KJ/m	MP°C	BP°C
Ag	(silver)		10.50	961.9	2212	4.29	Ag_2O	–1.85	230	
Al	(aluminum)	70	2.70	660.4	2467	2.36	Al_2O_3	–96.44	2072	2980
Au	(gold)		19.32	1064	2807	3.19				
Be	(beryllium)	29	1.85	1278	2970	2.18				
B	(boron)		2.34	2100	2550(s*)	0.32				
Cd	(cadmium)		8.65	321	765	≈0.9				
C	(carbon)		1.8–2.3	3550	3367(s*)	0.01–26				
	(diamond)	120+	3.15–3.53							
Cr	(chromium)		7.19	1843	2672	0.97	Cr_2O_3	–65.55	2266	4000
Co	(cobalt)	22.8	8.9	1495	2870	1.05				
Cu	(copper)	119	8.96	1083	2567	4.03	CuO	–9.00	1326	
							Cu_2O	–9.68	1235	1800
Fe	(iron)	207	7.87	1535	2750	0.87	FeO	–15.59	1369	
							Fe_3O_4	–65.04	1594	
							Fe_2O_3	–47.73	1565	
Mg	(magnesium)	45.5	1.74	648.4	1090	1.57	MgO	–34.39	2852	3600
Mn	(manganese)		7.3	1244	1962	0.08				
Mo	(molybdenum)	350	10.22	2617	4612	1.39				
Ni	(nickel)	207	8.9	1453	2732	0.94	NiO	–13.76	1984	
Pb	(lead)	14	11.35	327.5	1740	0.36	PbO	–12.60	886	
Si	(silicon)		2.33	1429	2355	1.68	SiO_2	–50.14	1723	2230
Sn	(tin)		5.75	231.97	2270	0.5–0.7	SnO_2	–16.37	1630	1800(s*)
Ta	(tantalum)	189	16.65	2996	5425	0.57	Ta_2O_5	–117.61	1872	
Ti	(titanium)	116	4.54	1690	3278	0.23	TiO_2	–28.84	1825	
V	(vanadium)		6.11	1900	3380	0.31				
W	(tungsten)	434	19.3	3598	5660	1.77	WO_3	–48.01	1473	
Zn	(zinc)		7.13	419.6	907	1.17	ZnO	–20.21	1975	
Zr	(zirconium)	84	6.5	1836	4377	0.23	ZrO2	–62.76	2715	

* sublimes

Now ρ_o depends on the pressure and temperature and N_o depends on the gas. Finally, from mean free path considerations and the fact that at 1 Torr (≈1.33 × 10^2 Pa) there are 3.54 × 10^{19} molecules in a liter of gas, we get:

$$\rho_o = \frac{3.5\times10^{22}\,P}{\sqrt{MT}}$$

P = pressure in Torr
T = temperature in degrees Kelvin
M = molecular weight (big molecules move more slowly)

Results for N_2 at 250°F (121°C or 394°K) and 10^{-6} Torr (1.4 × 10^{-9} atmos. or 1.33 × 10^{-4} Pa) are shown in the first two columns of Table 3.3:

Table 3.3 Time Required for Monolayers of N_2 to Adsorb on Glass

% covered	t, sec in 1.33 × 10^{-4} Pa at 121°C	t, sec. in Earth atmosphere (0.1 MPa) at 20°C	
25	0.8	3.2 × 10^{-8}	
50	1.7	6.8 × 10^{-8}	(The cross-sectional area of
75	3.5	14 × 10^{-8}	a molecule of nitrogen is
90	6.0	24 × 10^{-8}	about 16.2 Å^2 so about
95	7.5	30 × 10^{-8}	8.1 × 10^{14} molecules can be
99	12.0	48 × 10^{-8}	placed on an area 1 mm^2)

We may further estimate the time to adsorb gases at atmospheric pressure and temperature (where condensation of molecules is impeded somewhat by reevaporating molecules). This reduces the bombardment rate by about 1 order of 10, and at 20°C the bombardment rate is further reduced from that at 121°C by about 1/3 (altogether a factor of 1.4 × 10^{-9} × 10 × 3). The results are shown in the third column in Table 3.3. It may be seen that 90% coverage of one surface is achieved in 1/4 µs, a very short time!

The second and successive layers adsorb more slowly depending on many factors. Water adsorbs up to 2 to 3 monolayers on absolutely clean surfaces: contaminants, such as fatty acids, attract very many more layers than 2 or 3.

Oxidation begins as quickly as adsorption occurs. The rate of oxidation quickly slows down because of the time required either for oxygen to diffuse through oxide to get to the oxide/metal interface or for iron ions to migrate out to the surface of the oxide where they can join with oxygen.

Some experiments were done with annealed 1020 steel in a vacuum chamber, controlled to various pressures. The steel was fractured in tension, the two ends were held apart for various times, touched together, and then pulled apart again to measure readhesion strength. During the touching together, the relative amount of transmission of vibration at ultrasonic frequency through the partially reattached fractured ends was measured. The amount of exposure to gas bombardment is given in terms of Torr-sec. (time and pressure in the chamber).

Exposure, Torr-sec.	Relative adhesion	% of gas free surface	Ultrasonic transmission
10^{-6}	1	>95	>0.95
10^{-5}	0.95	≈50	≈0.9
10^{-4}	0.7	≈28	≈0.8
10^{-3}	0.4	≈7	≈0.5
10^{-2}	0.05	≈0	≈0.3

After the experiment with 10^{-2} Torr-sec exposure, a force was applied to press the fractured ends together. A load of 0.5 kN on a specimen of 10 mm diameter restored the ultrasonic transmission to the level of the experiment done at 10^{-4} Torr-sec, and a load of 1 kN restores it to the level of the 10^{-5} Torr-sec experiment. The adsorbed gas appeared to act as a liquid in these experiments.

(See Problem Set question 3 c.)

CHAPTER 4

Solid Surfaces

SURFACES ARE VERY DIFFICULT TO REPRESENT PROPERLY IN TRIBOLOGICAL MODELS. OUR INSTRUMENTS ARE TOO CRUDE, OUR MATHEMATICS TOO SIMPLE, AND OUR RESEARCH BUDGETS TOO SMALL TO CHARACTERIZE THEM WELL.

TECHNOLOGICAL SURFACE MAKING

Surfaces are produced in a wide variety of ways, and each process produces its peculiar roughness, subsurface damage, and residual stress. Several processes will be described.

Cutting: One of the more common surface making processes is done with a hard tool on metals (which are usually softer than 40 Rc) in lathes, milling machines, and drilling machines. (Steels as hard as 60 Rc can be cut with very hard tools such as cubic boron nitride.) Material removal in a lathe is done by a tool moving (usually) from right to left while a cylinder rotates. The finished surface is somewhat like a very shallow screw thread, depending on the rate of tool motion and the shape at the end of the tool. For some uses, this roughness of the cylinder along its length, i.e., across the screw threads or feed marks, adequately characterizes the surface. For many uses, however, the roughness in the direction of cutting is more important, particularly when using tools designed to minimize the feed marks.

The mechanics of cutting is usually represented as being done with a perfectly sharp tool edge. Such tools are difficult to make as is seen in the difficulty in getting very sharp points for use in scanning tunnel microscopes or field ion microscopes. Rather, practical tool "edges" can best be represented as being rounded, with radii, R, in the range of 2 to 40 μm. These dimensions are equivalent to 7000 to 60,000 atoms.

The cutting action of conventional tools can best be visualized by observing the cutting of fairly brittle metal such as molybdenum. Figure 4.1 is a sketch of a cutting process.

As the tool advances against the material to be removed it exerts a stress upon the material ahead of it. In a brittle material a crack initiates at some point where

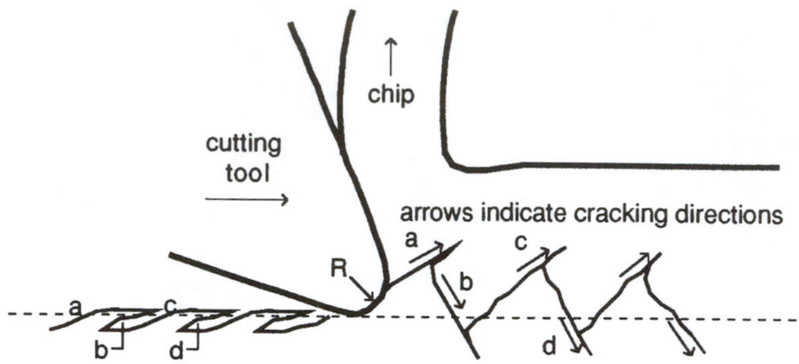

Figure 4.1 Sketch of the mechanics of cutting a brittle material.

the strength of the material is first reached and propagates along pathway "a." As the tool advances it imposes a changing stress field upon the material ahead of it until crack "a" has insufficient tensile stress to advance further. With further movement of the tool, the chip bends, exerting a tensile stress such that crack "b" initiates and propagates downward. This crack also moves into a diminishing stress field and stops. The stress field changes such that a new crack, "c," begins and propagates as shown. (Figure 4.1 shows a stationary tool but an advancing sequence of cracks.) The region below the cracks shows the shape of surface left by the crack sequences, which the heel of the tool alters further.

The sliding of the heel of the tool over a newly formed metal is a particularly severe form of sliding, producing very high friction. The tool burnishes the surface, pushing high regions downward, which causes valleys to rise by plastic flow. It shears the high regions so that tongues of metal become laps and folds lying over the lower regions. The result is a very severely deformed surface region that is particularly vulnerable to corrosion. This severe deformation extends about 5R to 10R into the surface. The surface is rough, but the laps and folds are relatively easily removed by later sliding. This is one reason why new surfaces wear faster during first use and why surfaces need to be broken in.

The above illustration uses brittle properties of material initially to explain how cracks propagate ahead of the tool but suggests plastic behavior under the heel of the tool. The latter is reasonable in brittle material because the material under the tool has large compressive stress components imposed.

Initially ductile material does not fracture in the manner shown in Figure 4.1, but a wandering pattern of shear is seen, followed by a finer pattern of ductile fracture planes. Fracture is likely to follow the interfaces between two phases so that the resulting surface topography will be affected by the sizes of grain and phase regions. Burnishing by the heel of the tool produces the same effect as described above.

The burnishing action is severe, resulting in a hardening of the surface layer. Strains of $\varepsilon \approx 3$ (and as high as $\varepsilon \approx 10$) can be inferred from hardness measurements. (An $\varepsilon=2$ can be achieved by stretching a mm-gage length of a tensile specimen to 14.8 mm and an $\varepsilon=10$ by stretching to 22 m.)

Rolling: Rolled sheet, plate, bar, et al., may be processed hot or cold. Hot rolling of metal is done at temperatures well above the recrystallization temperature and usually results in a surface covered by oxide and pock marks where oxides had been pressed into the metal and then fallen off. Cold rolling is usually done after thick scales of oxides are pickled off in an acid. It produces a smoother surface. There is some slip between rollers and sheet, which roughens the sheet surface, but this effect can be reduced by good lubrication.

Extrusion and Drawing: These processes can also be done hot or cold. The effect of oxides is the same as in rolling although the billets for extrusion and drawing are often heated in nonoxidizing atmospheres to reduce these effects. In any case, sliding of the deforming metal, polymer, and unsintered ceramic materials against hard dies (usually steel) will produce very rough surfaces unless the process is well lubricated.

Most cold-forming processes leave the surface of the processed part strained more (in shear) than the substrate has been strained. This produces surface hardening, but more important it produces compressive residual stresses in the surface with tensile residual stresses in the deeper substrate. (See the section titled *Residual Stress* in Chapter 2.)

Electrospark Erosion: This process (applicable mostly to metals) melts a small region of the surface and washes some molten metal away. The final surface roughness depends on the size of the "sparks" and the spacing between sparks if the electrode is moving. Just below the melt region the metal goes through a cycle of heating and cooling, leaving that region in a state of tensile residual stress. (See *Residual Stress,* Chapter 2.)

Grinding and Other Abrasive Operations: Removal of material by abrasive operations involves the same mechanics as in cutting with a hard tool. The major difference is the scale (size) of damage and plastic working. The abrasive particles (grit) in grinding wheels, hones, and abrasive paper are small but rounded primarily, and they produce grooves on surfaces. The abrasive particles cut (remove) very little material but they plastically deform the surface severely, as may be seen by the fact that abrasive operations require between 5 and 10 times more energy to remove a unit of material than do operations using a hard tool. Abrasive operations leave surfaces somewhat rough and severely cold-worked with residual stresses. Cold operations produce compressive residual stresses, but high severity grinding can produce tensile residual stresses.

(See Problem Set questions 4 a and b.)

RESIDUAL STRESSES IN PROCESSED SURFACES

Fracture, cutting, grinding, and polishing of ductile materials severely plastically deforms the surface layers, probably also producing a multitude of cracks extending into the solid. In cutting and grinding, the deformation comes from the fact that the cutting edges of tools and abrasive particles are rounded rather than perfectly sharp.

Localized plastic flow produces compressive stresses. Localized heating and cooling, as in grinding, can produce tensile stresses. An example of the intensity of these stresses can be seen in Figure 4.2.

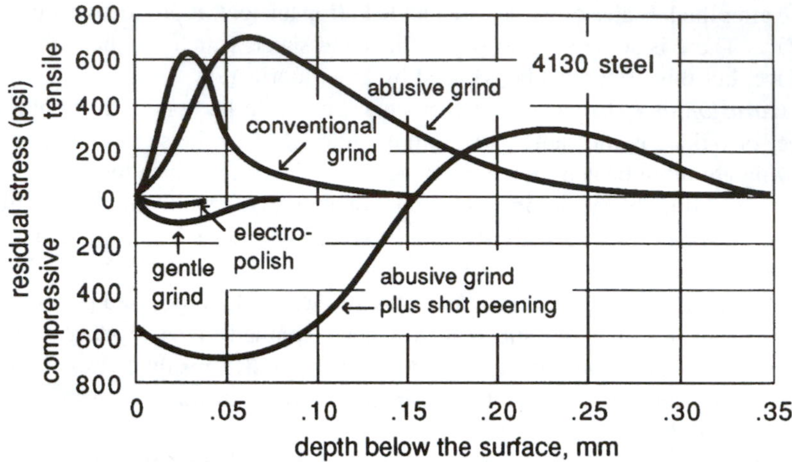

Figure 4.2 Residual stresses after various grinding operations upon 4130 steel. (Adapted from Koster, W.P., International Conference on Surface Technology, May 1973, Carnegie Mellon University, Society of Manufacturing Engineers, Dearborn, MI, 1973.)

Grinding conditions	Gentle	Conventional	Abusive
Wheel type	A46HY	A46KY	A46MY
Wheel speed m/s	10.2	30.5	30.5
Downfeed, mm/pass	<0.03	0.03	0.05
Grinding fluid	Sulfurized oil	Soluble oil	Dry

The extent of surface deformation is seen in polishing for the purpose of metallographic examination. The crystallographic structure of the metal is hidden by a layer of severely deformed metal. The structure of polished surfaces was studied by Sir George Beilby.[2] He found, by X-ray diffraction, that no crystalline structure appeared in the polished surface. He therefore suggested that this layer might be amorphous, and it became known as the Beilby Layer. Later work showed that this layer consists of very fine crystallites probably including embedded polishing compound and reaction products, and is not amorphous at all. Its thickness is defined by the process used to form it.

Beneath the very severely deformed region are gradations of less deformed material. These states of deformation are illustrated at the end of this section. Above the solid surface yet another phenomenon occurs, namely, oxidation and adsorption.

(See Problem Set question 4 c.)

ROUGHNESS OF SURFACES

The roughness of surfaces is expressed as the height of the small irregularities or asperities on the surfaces. The practical range of roughness of commercial surfaces is given in Table 4.1.

Table 4.1 Practical Range of Roughness of Commercial Surfaces, Units are R_a (1 μ in = 10^{-6} in)

Float glass (solidified while floating on molten tin or other metal)	1 nm, or (0.04 μ in)
Polished plate glass and highly polished metal	1.8 nm, or (0.07 μ in)
Commercial polishing, and the products cast in such polished molds	0.1 μm (4 μ in) to 0.01 μm (0.4 μ in)
Commercial grinding	0.25 μm (10 μ in) to 0.025 μm (1 μ in)
Good machined surfaces (cut by hard tools)	2.5 μm (100 μ in) to 0.25 μm (10 μ in)
Rolled and drawn surfaces	10 μm (400 μ in) to 1.0 μm (40 μ in)
Sand cast surfaces	25 μm (1000 μ in) to 2.5 μm (100 μ in)

It is obvious that all of the roughnesses described above are large compared with nm units (atoms are of the order of 0.3 nm apart). Atomic models lose their significance in the face of such great roughnesses. However, since the majority of surfaces that come into contact with each other have relatively rough surfaces, we shall spend most of our time with such surfaces. (The Appendix to Chapter 12 contains a section on *Surface Roughness Measurement*.)

FINAL CONCLUSIONS ON SURFACE LAYERS

Surfaces are quite complicated. From various sources we may estimate the thickness of various layers on a tool-cut surface that has been exposed to atmosphere for a day or two. (Iron oxide becomes ≈ 25Å thick in 10 minutes in a pressure of 80×10^{-5} Torr or ≈ 0.1 Pa. This indicates the early growth rate of oxides. In contrast, Figure 4.3 shows oxide thickness in the 3 to 15 nm range, which takes hours to days.)

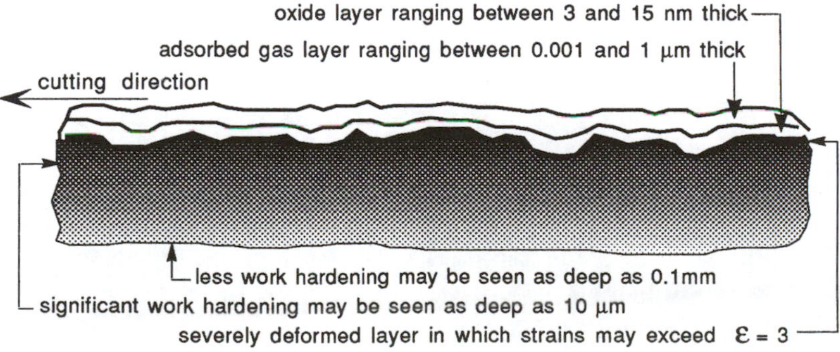
oxide layer ranging between 3 and 15 nm thick
adsorbed gas layer ranging between 0.001 and 1 μm thick
cutting direction
less work hardening may be seen as deep as 0.1mm
significant work hardening may be seen as deep as 10 μm
severely deformed layer in which strains may exceed ε = 3

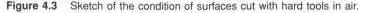

Figure 4.3 Sketch of the condition of surfaces cut with hard tools in air.

In grinding, the highly strained solid layers may be one tenth as thick as those shown in Figure 4.3, and in fine polishing these layers may be even thinner.

REFERENCES

1. Koster, W.P., International Conference on Surface Technology, May 1973, Carnegie Mellon University, Society of Manufacturing Engineers, Dearborn, MI, 1973.
2. Bowden F.P. and Tabor, D., *Friction and Lubrication of Solids*, Oxford University Press, Oxford, U.K., 1954.

CHAPTER 5

Contact of Nonconforming Surfaces and Temperature Rise on Sliding Surfaces

MECHANICS IS STRONGLY BASED ON ASSUMPTIONS OF IDEAL SURFACE SHAPES AND IDEAL MATERIALS BECAUSE REAL SURFACE TOPOGRAPHY AND MATERIALS ARE VERY DIFFICULT TO CHARACTERIZE. BETTER DETAIL IS NOT NEEDED TO SOLVE MOST PROBLEMS, BUT TRIBOLOGY PROBLEMS DO REQUIRE MORE DETAIL.

CONTACT MECHANICS OF NORMAL LOADING[1,2]

Surfaces are usually rough (have asperities on them) so contact between them can only occur at a limited number of points. The pressure on those points is therefore very high. We can make some assumptions about this area of contact if we make some assumptions about the nature of asperities. The point of doing so is to develop the basis for discussions on real area of contact and temperature rise on sliding surfaces. These quantities were prominent in early research and in the development of models for friction and wear.

In general, contact involves both elastic and plastic ranges of strain within and beneath asperities. Thus asperities should be modeled in such a way that both elastic and plastic deformation zones may be seen. The cone-shaped asperity is therefore excluded because a finite load on a point will cause infinite stress, except for the possible case of contact on the sides of cones. The sphere is a much-used model. Equations are available in many forms, the most convenient of which give the size of the contact region and the stress distribution in that region. The equations were derived by Heinrich Hertz in 1881 (at the age of 28!). The equations often contain Young's Modulus (E) and the Poisson ratio (υ) in a bulky form, which will be simplified as follows:

$$q_o = 0.578\left(\frac{W}{(RN)^2}\right)$$

and

$$\frac{1}{R} = \frac{1}{R_1} + \frac{1}{R_2}$$

where subscripts refer to body 1 and body 2.

Hertz assumed a semi-elliptical stress distribution between the bodies leading to the following equations for two solids contacting each with load W applied, and with π and exponents on numbers all worked out:

For two spheres the maximum contact stress is:

$$q_o = 0.578\left(\frac{W}{(RN)^2}\right)^{1/3} \quad \textbf{(1)}$$

and the radius of contact is:

$$a = 0.91(WNR)^{1/3} \quad \textbf{(2)}$$

For parallel cylinders of length L, the maximum contact stress is:

$$q_o = 0.564\left(\frac{W}{RNL}\right)^{1/2}$$

and the region of contact has the half-width of:

$$b = 1.13\left(\frac{WNR}{L}\right)^{1/2}$$

where one body is flat, R_2 is ∞. For a sphere in a socket or a shaft in a bearing, R_2 is negative.

The influence of υ is relatively small in these equations. The full range of υ for commercial materials is from 0.05 for beryllium to 0.5 for rubber. (Most metals have $\upsilon \approx 0.3$.) Over this range of υ, the full range of calculated values of "a" for the sphere is 9%; the range on q_o for the sphere is 18%; and the range on both quantities for the cylinder is 15%.

The average (mean) pressure of spherical contact is $p_m = W/\pi a^2$. For a semi-elliptical pressure distribution over the area of contact the maximum pressure q_o is $(3/2)p_m$.

Another equation inserted here because of its similarity to the above group gives the distance between the centers of two spheres that come together when loaded:

$$B = 1.21\left(\frac{(WN)^2}{R}\right)^{1/3}$$

Hertz also provided equations for the stress distribution below the center of contact as shown in Figure 5.1, for a sphere on a flat plate. The highest shear stress occurs at the point of greatest difference between σ_z and σ_r. That turns out to be a depth of 0.5a. (For two flat plates, 0.5a is very large.) The maximum calculated shear stress, $(\sigma_z - \sigma_r) = \tau_{zr} = 0.47\ p_m = 0.31\ q_o$. In the simplest view, plastic flow (shearing) occurs when $\tau_{zr} = Y/2$ (Y= the tensile yield strength), then plastic flow first occurs under the condition:

$$0.47\ p_m = 0.5\ Y \quad \text{or} \quad p_m \approx 1.1\ Y \tag{3}$$

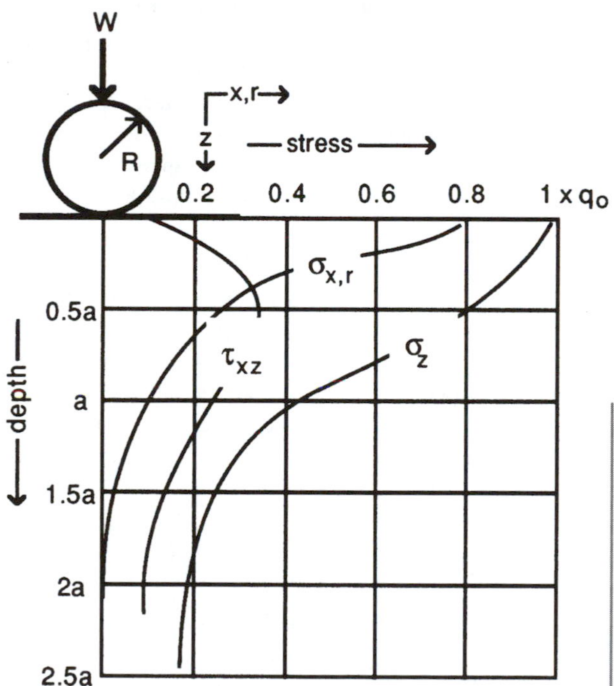

Figure 5.1 Stress state along the vertical axis under a ball pressed against a flat plate.

With continued loading of the ball, the small plastically deformed region grows and the mean pressure increases. Experimentally the mean pressure, p_m, has been found to approach 2.8 Y as the load, W, increases as shown at the left in Figure 5.2. (For work hardening metals the value of Y is taken as that at the edge of the indentation at any instant.)

With the onset of plastic flow the elliptic stress distribution assumed by Hertz no longer applies. A. J. Ishlinsky has published an approximate stress distribution

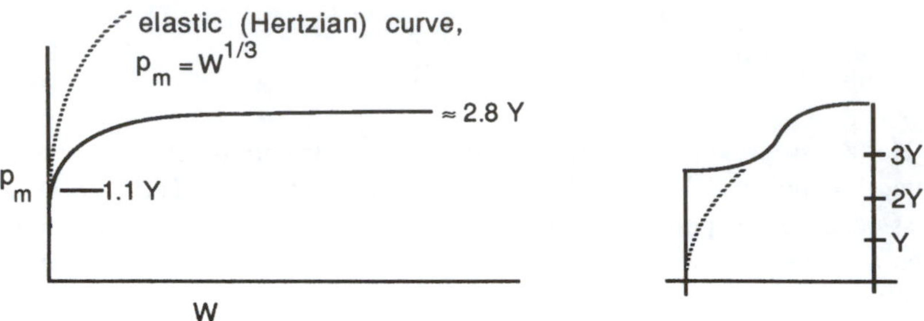

Figure 5.2 Stress state for loading that produces plastic flow.

for the ideal, plastic case as shown in the right panel in Figure 5.2. The elliptic distribution is dotted-in to show similarity. This means that p_m is no longer equal to $2/3q_o$. An exact stress distribution was difficult to derive because of the tedious nature of locating the boundary between elastic fields and plastic fields. Stress fields of all types can now be mapped by Finite Element methods, with which it is equally easy to use any of the available yield criteria. (Conclusions reached by simple methods are adequate for understanding, and often yield results with uncertainties no greater than the uncertainties in given values of mechanical properties of materials. For example, the Young's Modulus for steel ranges from ≈ 182 to 233 GPa.)

Deep indentation of a sphere into a flat plate is commonly done in hardness testing such as with the Brinell system or Rockwell "b" system. One important conclusion we may reach is that in hardness testing the yield strength of the indenter must be at least three times that of the metal being indented.

An elastic stress field in a flat plate, pressed by a *cylinder* is shown in Figure 5.3, showing the magnitude and direction of maximum shear stresses in terms of q_o. Note that the depth at which the maximum shear stress exists is 0.78b.

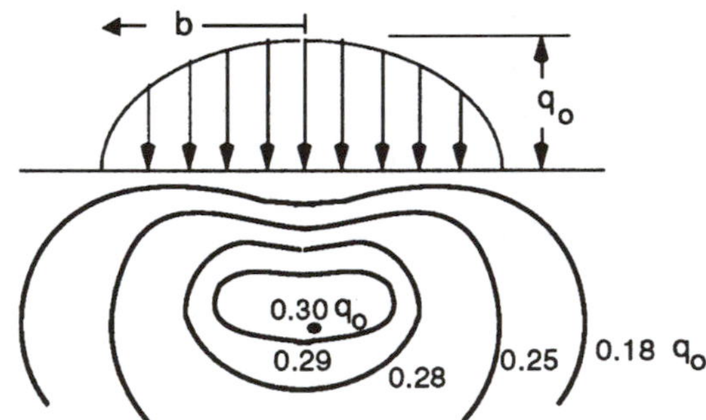

Figure 5.3 Shear stress contour in a flat plate indented by a cylinder, in terms of q_o.

The stress fields change due to friction as when a force is applied to a sphere or cylinder to slide it. Shear stress contours in a flat plate indented by a cylinder are shown in Figure 5.4 for the case of a ratio of shear force to normal force of

0.2. This stress state exists when sliding occurs and the coefficient of friction is 0.2.

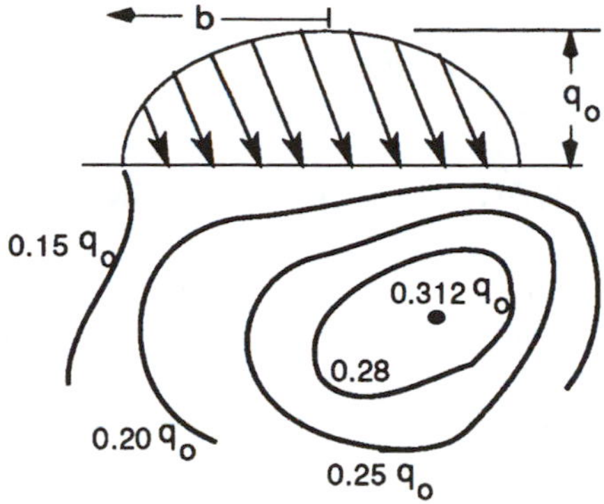

Figure 5.4 Shear stress contours for cylinder indenting a flat plate with a ratio of shear force to normal force of 0.2.

For a ratio of shear force to normal force (i.e., a coefficient of friction) of about 0.3 or greater, the point of maximum shear stress emerges to the surface.

Figure 5.2 would suggest that stresses $p_m \approx 3Y$ must be applied to continue plastic flow. This seems to be inconsistent with the stress states in tensile tests. The reason is that the small volume of plastically deforming metal is constrained by the large surrounding elastic field. If adjacent asperities are very close, they no longer have independent elastic stress fields supporting them.

(See Problem Set question 5 a.)

RECOVERY UPON UNLOADING

When a sphere presses into a flat plate the contacting regions of the two bodies conform perfectly. If the load is removed, separation of the sphere from the plate begins from the outer edge of contact and moves inward.

If only elastic deformation has occurred, both bodies return to their original shapes. If, for example, the flat plate had plastically deformed locally, upon removal of the sphere a dent is left in the plate. This indicates clearly that much of the stress field shown in Figure 5.1 remains in, or resides in, the plate.

If a load is applied such that $p_m \leq 2.15Y$, upon removal of the load the elastic stress field does not apply sufficient force upon the plastically deformed volume to cause reverse plastic flow (shear). Subsequent repeated loading and removal of the same load produces only elastic strain cycling in the flat plate.

If a load is applied such that $p_m > 2.15$ Y, upon removal of the load the elastic stress field causes reverse plastic flow in the plastically deformed volume. Repeat loading causes plastic strain cycling. With each cycle the sphere sinks a little farther into the flat plate (to some limit).

Wheels on rails produce the same effect as does a sphere on a flat plate. Plastic strain progression by a succession of highly loaded wheels makes a layer of rail shear forward relative to the deeper substrate (eventually resulting in fatigue failure).

(See Problem Set question 5 b.)

ADHESIVE CONTACT OF LOCALLY CONTACTING BODIES[1]

In the previous section the loading of a sphere against a flat plate was discussed. The same would apply to the pressing of a soft rubber ball against a flat plate. The previous discussion applied to the case of no sensible adhesion between the two bodies. Releasing the load allows each body to deform out of conformity to each other, and separate. The driving force is supplied by relaxation of the strain energy in the two substrates which was imposed by applying the load.

When the two bodies stick together upon loading, a new stress state prevails upon unloading. Take the case of a sphere pressing into a flat plate and restrict ourselves to the elastic case. There is a contact area of radius "a" as given before. Now suppose the two surfaces adhere over the contact area. If both bodies have the same υ and E, the contour of the contact region will not be affected by releasing the load (because of adhesion), and yet releasing the load is like applying a reverse load, W′. Applying W′ to an unchanging surface contour produces the same stress distribution (though reversed) as pressing a rigid (sharp cornered) circular cylinder against a flat plate. This produces a pressure distribution at distances, "x," from the center as given in Equation 4.

$$P' = \frac{W'}{2\pi a^2 \sqrt{1 - \frac{r^2}{a^2}}} \quad \textbf{(4)}$$

Note that the stress at the periphery of the contact area is infinite whether added to the elliptical contact pressure distribution or not.

This analysis uses unrealistic material properties, but it shows clearly the source of the tearing force. In the usual case the high stress at the edge of contact is alleviated but not eliminated by plastic flow. Thus, if the asperities stretch plastically at the periphery, contact is maintained and more force will be necessary to separate the parts.

A practical illustration of this effect may be seen using a rubber ball on a plate. When viewing through the glass plate, the area of contact is seen to vary with applied load. Cover the glass plate with a thin layer of a very sticky substance. Now press the ball against the flat plate and suddenly release the load. The ball recovers its shape slowly. Strands of the sticky substance can be seen to bridge the gap where once the bodies were in contact. After some time a small region of adhesion remains. Metals behave the same way, only much more quickly and on a microscopic scale. (See the section titled *Adhesion* in Chapter 3.)

AREA OF CONTACT[2]

Studies of contact stress were common in the 1930s when research focused strongly on deciding between the adhesion theory of friction and the interlocking theory of friction. It was thought that the question could be resolved by knowing the amount of real contact area (sum of the tiny asperity contact areas) between contacting and sliding bodies. That there is a large difference between real and apparent area of contact had been known for some time, particularly by people who had no concern for theories of friction, however. As a result, most people understand why the flow of heat and electricity through contacting surfaces is enhanced by increasing contact pressure.

Apparent (or nominal) area of contact is that which is usually measured, such as between a tire and the road surface or calculated for the case of a large sphere on a rough flat plate, by equations of elasticity as in the previous section. Real area of contact occurs between the asperities of surfaces in contact. *If* all contacting asperities were in the fully developed plastic state, the contact pressure in them all would be about 2.8 Y, or for convenience $\approx$ 3Y. Thus, the area of contact $A_r \approx W/3Y$. For 1020 steel with the yield strength Y = 150,000 psi (1 GPa), a 1-inch cube pressed against a flat plate of steel with a load W produces a real contact area of A_r:

W	A_r
10,000 lb	1/15 in^2
100 lb	1/1500 in^2
1 lb	1/150,000 in^2

A person has the strength to indent a steel anvil! For a 1/2-inch ball pressed with 10 lb., $q_o \approx 10^5$ psi, which is about the yield strength of anvil steel.

Note that all asperities are assumed to be fully plastic in the calculation above. Actually, some of them will be elastically deformed only, so that the real area of contact will be larger than calculated above. However, well over 90% of the load is carried on fully developed plastically deformed asperities.[3]

A great number of methods have been attempted to measure real area of contact, but all methods have shortcomings. Five methods and limitations are listed:

1. Two large model surfaces with asperities greater than 1 inch in radius, one covered with ink which transfers to the other at points of contact. Acceptable simulation of microscopic asperities has not yet been achieved.
2. Electrical resistance method. This method is limited by surface oxides and by the fact that electrical constriction resistance is related to $\Sigma 1/a$ and not $\Sigma 1/a^2$ (discussed in a later section).
3. Adhesion and separation of sticky surfaces. In this method two clean metal surfaces in a vacuum are touched together with a small force and then pulled apart. The force to separate was thought to be W = 3YA. This method is limited

by elastic recovery when load is removed and by fracture of bonds that may extend beyond the contact region.

4. Optical method, interference, phase contrast, total internal reflectance, etc. With these methods it is difficult to resolve the thickness of the wedge of air outside of real contact area down to atomic units, which is the separation required to prevent adhesion.
5. Acoustic transmission through the contact region between two bodies, and again the measured area is related to $\sum a$ and not $\sum a^2$.

In the absence of good measurement methods, researchers have always inferred the area of contact from contact mechanics. To summarize the case of contact between a single pair of spheres:

1. Elastic case, $A \propto W^{2/3}$.
2. Plastic case, $A \propto W^1$.
3. Visco-elastic case, "A" changes with time of contact.

   In real systems consisting of complex arrays of asperities, the following conclusions have been reached, largely through experiments:
4. In metal systems, ranging from the annealed state to the fully hardened state, contact appears to produce large strain plastic flow. Thus, $A \propto W$. This simplifies matters greatly. Recall that we have considered hemispherical asperities for convenience. It happens that where we take asperities of conical or pyramidal shape against a flat plate $p_f \approx 3Y$ (p_f = flow pressure which is the yield strength in multiaxial deformation) for larger cone angles, and higher than 3Y for smaller cone angles. But by experiment $A \propto W$ for almost every conceivable metal surface, which probably indicates that asperities may be taken to be spherical in shape for purposes of analysis.
5. In most nonmetal systems contact appears to be nearer to elastic. For rubber, plastic, wood, textiles, etc. $A \propto W^n$ where $n \approx {}^2/_3$. For rock salt, glass, diamond, and other such brittle materials "n" may be nearer to 1 than ${}^2/_3$. Thus, these brittle materials appear to deform plastically. However, there may be another reason. Archard found mathematically that for:[4]

Single smooth sphere	$A \propto W^{2/3}$
Single sphere with first order* bumps	$A \propto W^{8/9}$
Single sphere with second order* bumps	$A \propto W^{26/27}$
Several spheres of different heights	$A \propto W^{4/5}$
Several spheres with first order* bumps	$A \propto W^{14/15}$
Several spheres with second order* bumps	$A \propto W^{44/45}$

* widely separated orders

Glass, diamond, etc. may have complex asperities unless cleaned or fire polished. On the other hand, the $n \approx 2/3$ for the other elastic solids mentioned may imply that asperities on these are relatively simple in nature, perhaps having a few first order bumps but not second order bumps.

These are elastic calculations and can be in error if the influence of close proximity of asperities is ignored. When plastic strain fields of closely spaced asperities overlap, several asperities act as one larger asperity.

(See Problem Set question 5 c.)

ELECTRICAL AND THERMAL RESISTANCE

Electrical resistance across a contact area is greater than the sum of the resistances of the elements, r_t, as shown in Figure 5.5. Holm reported the mathematical work of Maxwell which showed the need for a correction due to the constriction of the stream of current in the regions of r_2 and r_4.[1] Holm himself measured values quite carefully and found, for two large bodies joined by one bridge of radius r, $R = 1/(2a\,\lambda)$ where λ is the specific conductance of the metal. An oxide on each surface adds some resistance so that the total may be $R = 1/(2a\,\lambda) + 2\sigma/(\pi a^2)$ where σ is the resistance per unit area of the layer of oxide. In many cases, the oxide may be the chief cause of resistance.

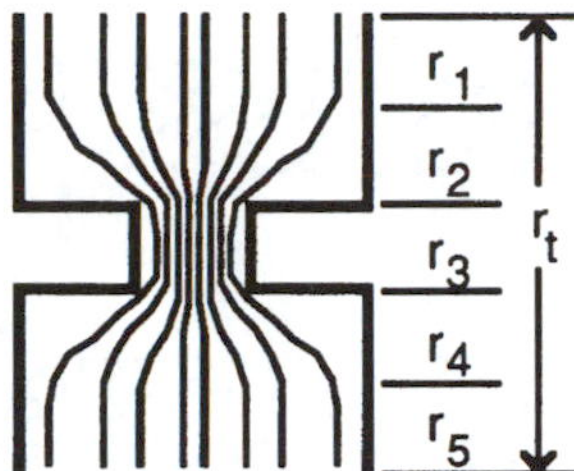

Figure 5.5 Summation of electrical resistance through a contact bridge.

dR/dt shows the rate of oxidation. (Electrical contact resistance has been used to measure A but the results have usually been ambiguous.) The resistance of a piece of a material may be calculated by $R = \rho L/A$ where ρ values are:

Material	Resistivity, ρ
Copper	1.75 μ-ohm-cm
Aluminum	2.83
Platinum	10
Iron	10
Marble	10^{11}
Porcelain	10^{14}
Glass	10^{14}
Hard rubber	10^{18}

SURFACE TEMPERATURE IN SLIDING CONTACT [5]

Frictional energy heats sliding bodies, which may produce a strong effect on local material properties, chemical reactivity of lubricants, oxidation rates, initiation of explosive reactions in unstable compounds, and the formation of sparks (dangerous in mines, particularly in an atmosphere of ≈7% methane in air, for example).

Calculation of heat transfer rates and temperature distribution is rather daunting because it involves so many dimensional units. The temperature rise on sliding surfaces is a particularly complicated problem, primarily because of its transient nature. Most tribologists would prefer to leave the topic to those who work in the field as a career, but sometimes it is necessary to estimate surface temperatures of sliding bodies in engineering practice.

The major concern among tribologists is to choose a useful equation from among the many available in the literature. Several of the more widely discussed will now be presented, as will a perspective on methods and accuracy of equations. The case of greatest interest in sliding is the pin-on-disk geometry. Assume a pin made of conducting material, surrounded by (perfect) insulation, and held by an infinite mass of very much higher thermal conductivity than the pin, as shown in Figure 5.6.

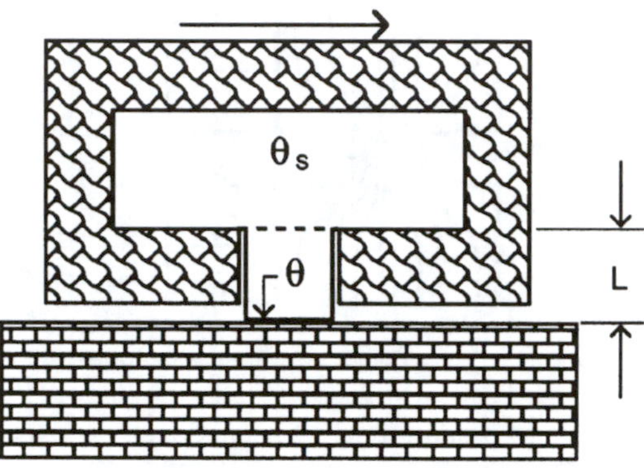

Figure 5.6 Sketch of a conducting material sliding over an insulating material.

The pin slides along a flat plate of a perfect insulator with zero heat capacity. That is, none of the frictional heat is conducted into the flat plate and no heat is required to heat the surface layers of the flat plate. Then all of the frictional energy is conducted along the length of the pin as shown in the sketch. After equilibrium is established, the average temperature of the sliding end of the pin can be calculated as $\theta = \alpha L \theta_s$, where α is the heat transfer coefficient, L is the length of the pin, and θ_s is the temperature of the heat sink.

Now assume the opposite case, i.e., a plate of *conducting* material upon which a pin slides and the pin is made of the perfectly insulating material with zero heat capacity. The simplest assumption in this case is that the temperature across the end of the pin is uniform. This is the assumption of the uniform heat flux or uniform heat input rate. If that heat source is stationary, then in the first instant the temperature distribution across the surface of the plate (assume the two-dimensional case) is as shown as the rectangular curve 0 in Figure 5.7.

After some time, heat will flow to the left and right and if the rate of heat input is just sufficient to maintain the same maximum temperature as for curve 0 in Figure 5.7, then the temperature gradient is shown as curve 1, then 2, etc., in the figure. However, if the heat source had been shut off after curve 0 then the temperature distribution would change as shown in curve 3.

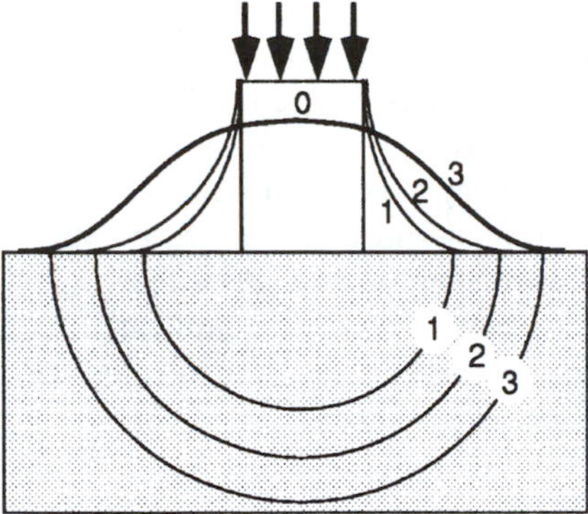

Figure 5.7 Temperature profile over a surface upon which heat is impinging. Rectangular distribution 0 exists for a very brief time after initiation of heating; distributions 1 and 2 exist after some time of heating; and distribution 3 exists after the heat source is removed.

If the heat source moves to the right, the surface material to the left cools by conduction of heat into the substrate, and the material to the right begins to heat. If the rate of heat input is equal to the rate of exposure to new surface times the amount of heat required to heat the material to the same temperature as before, the temperature distribution will be skewed as shown in Figure 5.8.

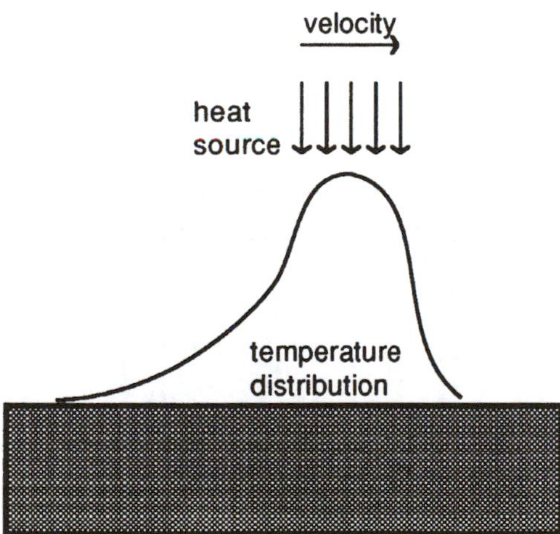

Figure 5.8 Temperature distribution on a surface from a moving source.

The maximum temperature will be *near* the rear edge of contact rather than *at* the edge because heat is transferred away from the heated region. Further, it may be seen that the higher the velocity of movement of the heat source relative to the thermal conductivity of the plate material, the nearer the maximum temperature will be to the rear edge of contact.

Now, each of the pins and the plates in Figures 5.7 and 5.8 have different temperature distributions on their surfaces. In practical sliding systems, neither the pin nor the plate are insulators, or are insulated, generally. For analysis, the temperature distributions over the region of apparent contact in each body are assumed to be the same, though not uniform. In other words, the mathematical solutions to each of the above ideal cases are combined, taking the contact temperature distributions on both surfaces to be the same. The complete solution of the pin-on-disk sliding problem is very complicated. Engineers have therefore found it convenient to present equations for average surface temperature over nominal contact areas for several special cases. For these equations, the symbols are given first:

$\theta_{ave.}$	=	difference between the average temperature on the sliding interface and temperature in the solids far removed from the sliding interface
V	=	velocity of movement of the heat source = sliding speed
μ	=	coefficient of friction
W	=	applied load
L	=	cross-sectional dimensions of the square pin
J	=	mechanical equivalent of heat
κ	=	thermal diffusivity = $k/\rho c_p$ where
k	=	thermal conductivity; k_1 for the plate and k_2 for the pin
ρ	=	density of the solid
c_p	=	specific heat of the solid
g	=	gravitational units, optional depending on units used elsewhere

Equations for two of these cases are:

where the sliding speed is small relative to the rate of heat flowing away from the contact area, and assuming no phase change, the surface temperature rise over ambient, θ, is:

$$\text{for } \frac{VL}{2\kappa} < 0.1, \qquad \theta_{ave.} = \frac{0.236\mu WV}{LJ(k_1 + k_2)} \tag{5}$$

in the case of high sliding speed and low heat flow rate:

$$\text{for } \frac{VL}{2\kappa} > 5, \quad \theta_{ave.} = \frac{0.266(\kappa_1)^{1/2}\mu WV(g)}{LJ[1.124\, k_2(\kappa_1)^{0.5} + k_1(LV)^{0.5}]} \tag{6}$$

Tabor derived a similar equation based on the form of the Holm equations for electrical (constricted) conductivity through an interface. He (as most others do) interposed a thin plate between asperities on two surfaces. The total frictional heat generated flows through asperity contact regions of radius a, into the two bodies, $Q = Q_1 + Q_2$. The quantity 4a is the Holm representation of contact area:

$$Q_1 = 4ak_1(\theta_{ave}) \quad \text{and} \quad Q_2 = 4ak_2(\theta_{ave})$$

$$\theta_{ave} = \frac{Q}{4a(k_1 + k_2)}$$

$$\text{now } Q = \text{heat input} = \frac{\mu WV}{J}$$

$$\text{so, } \theta_{ave} = \frac{\mu WV}{4aJ(k_1 + k_2)} \tag{7}$$

Equation 7 produces results about 6% higher than results from Equation 5.

From the above equations it would seem that the influence of speed and load can be expressed as:

$$(\theta_{ave}) \propto WV$$

This appears to conflict with the findings of Tabor in the 1950s.[2] In experiments where metal rubs on glass and the contact region is viewed through the glass, Tabor reported visible hot spots which he estimated to be about 10^{-4} inch in diameter and lasting about 10^{-4} sec. Three points emerge from this work:

1. Hot spots are never seen for metals with MP less than about 970°F to 1060°F. (Visible red heat begins in this temperature range.)
2. For metals with higher MP (than about 1000°F) hot spots are not seen until either V or W is increased.
3. The magnitudes of the factors V and W for the appearance of hot spots are related by $VW^{1/2}$ = const.

This apparent conflict may not be serious if we alter Tabor's equation for low V: For elastic contact

$$\frac{W}{a} = \frac{W^{2/3}}{1.1}\left(\frac{E}{r}\right)^{1/3} \quad \text{and} \quad \theta_{ave} = \frac{W^{2/3} V \mu g (E)^{1/3}}{4.4Jr(k_1 + k_2)} \tag{8}$$

(r = the radius of a spherical asperity)

For plastic contact

$$\frac{W}{a} = \sqrt{Wp_m \pi} \approx 3\sqrt{YW} \quad \text{and} \quad \theta_{ave} = \frac{3(WY)^{1/2} V \mu}{4J(k_1 + k_2)} \tag{9}$$

(Y = the yield strength of the material)

These equations apply to real contact area as distinct from the apparent contact area assumed in the previous equations. Since the exponent on W corresponds with experimental results, Tabor probably saw plastic behavior of asperities in his tests or else the material properties changed in a manner that appeared as if the effect of load should properly be represented as $W^{1/2}$. Recently other writers have suggested the need to account for thermal softening of the surface.

Whereas Tabor's equations apply over real contact areas, they apply to low values of VL/κ. These equations do not distinguish between pin or flat material, which is of little consequence at low VL/κ anyway.

Assuming the Tabor equations apply reasonably well to metals, what order of V causes melting? Calculations show the following critical sliding speed for a 1/8-diameter cylinder end of various metals on steel with a 100 gram load (≈25 psi) applied:

Gallium	100 f.p.m.
Lead	100 f.p.m.
Constantan	800 f.p.m.
Copper	60,000 f.p.m. (600 mph)

From these data it would appear that airplane brakes of alternate plates of steel and chromium copper are safe. The landing speed of a passenger airplane is about 150 mph and brakes slide at about 1/2 ground speed, or ≈ 75 mph. (Brake discs and miscellaneous associated parts on a Boeing 747 cost $25,000 per wheel, and on a 707 they cost $10,000. Metal brake disks last 20 to 40 landings depending on the amount of reverse thrusting used to aid braking, or one aborted take-off. An aborted take-off of a 707 costs the airline at least $25,000 in passenger handling plus the cost to repair the cause of abort, at 1980 prices. Carbon brakes are now more common and last much longer than metal brakes.)

(See Problem Set question 5 d).

COMPARISON OF EQUATIONS 5 THROUGH 9

Both Equation 5 and 6 are plotted as straight lines on log–log coordinates, but each has a different slope. The slope of Equation 5 is 1 (45°), whereas the slope of Equation 6 varies with the magnitude of the parameters used. These equations are plotted in Figure 5.9 for a copper pin of L = 0.63 cm (1/4″) pressing on a copper plate with a load of 22,700 grams (50 lb.). Note that there is a blend region between the two equations, and note also that a single equation for the full range of sliding speed shown in Figure 5.9 would be very complicated.

Equations 5 and 7 show nearly the same results for stainless steel, but Equations 8 and 9 show rather different results. Recall that Equations 5, 6, and 7 represent the average temperature rise in the *nominal* area of contact, whereas Equations 8 and 9 apply to the *real* areas of asperity contact and is the *flash temperature* that we read of in some papers. The flash temperature for elastic contact (Equation 8) is much higher than that for plastic contact (Equation 9) because no elastic limit (i.e., yield point) is imposed upon contact pressure. Thus there are few, but very hot, points of asperity contact.

All equations are shown intersecting a vertical line at the arbitrarily selected sliding speed of 1.3 m/s, which is walking speed (≈ 250 f/m or 3 mph). This sliding speed is near that at which the transition occurs between Equations 5 and 6 for copper sliding on copper. Restriction to this area also yields the impracticably small values of temperature rise seen in Figure 5.9.

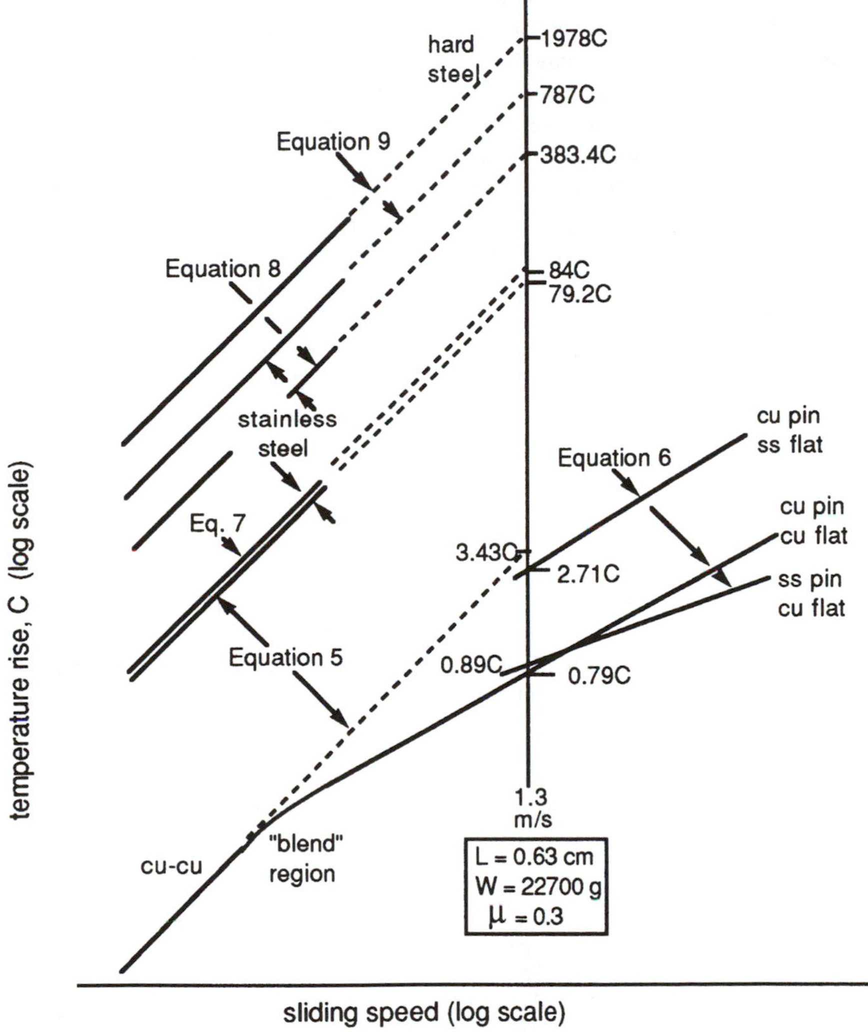

Figure 5.9 Plot of Equations 5–9 on log–log axes, temperature rise versus sliding speed.

It is seen that different assumptions produce fairly large differences in results and that for higher sliding speeds it is necessary to know which of a dissimilar pair is the pin or the disk. Further, it may be inferred that for other contacting pairs, completely different equations are required, such as for cams and followers, for gear teeth, and for shafts that whirl in the bearings.

TEMPERATURE MEASUREMENT

The measurement of *surface* temperature has usually been attempted with either the embedded thermocouple or with the Herbert-Gottwien (contact between dissimilar metals) thermocouple. The results hardly ever agree. The embedded thermocouple cannot be placed closely enough to the surface to read real and instantaneous temperature, certainly not of asperities. The dynamic thermocouple measures the electromotive force (emf) from many points of microscopic contact simultaneously, and the final result will be a value probably below the average of the surface temperature of the points. Errors as large as 100°C are highly likely.

Surface temperatures are also measured by radiation detectors. Again these devices measure the average temperature over a finite spot diameter. Size depends on the detector. For opaque materials the measurements may be made after the sliders have separated, with some loss of instantaneous data. Where one of the surfaces is transparent, the radiation that passes through can provide a good approximation of the real temperature. All of these methods require extensive calibration.

REFERENCES

1. Johnson, K.L., *Contact Mechanics,* Cambridge University Press, Cambridge, U.K., 1985.
2. Bowden, F.P. and Tabor, D., *Friction and Lubrication of Solids*, Oxford University Press, Oxford, U.K., Vol. 1, 1954, Vol. 2, 1964.
3. Greenwood J. A. and Williamson, J.B.P. *Proc. Roy. Soc.,* (London), A295, 300, 1966.
4. Archard, J.F., *Proc. Roy. Soc.,* (London), A 243, 190, 1958.
5. Dorinson, A. and Ludema, K.C, *Chemistry and Mechanics in Lubrication*, Elsevier, Lausanne, 1985, Chapter 15.

CHAPTER 6

Friction

FRICTION IS A FORCE THAT RESISTS SLIDING. IT IS DESCRIBED IN TERMS OF A COEFFICIENT, AND IS ALMOST ALWAYS ASSUMED TO BE CONSTANT AND SPECIFIC TO EACH MATERIAL. THESE SIMPLE CONCEPTS OBSCURE THE CAUSES OF MANY PROBLEMS IN SLIDING SYSTEMS, PARTICULARLY IN THOSE THAT VIBRATE.

CLASSIFICATION OF FRICTIONAL CONTACTS

Some surfaces are expected to slide and others are not. Four categories within which high or low friction may be desirable are given below.

1. *Force transmitting components* that are expected to operate without interface displacement. Examples fall into the following two classes:

a. Drive surfaces or traction surfaces such as power belts, shoes on the floor, and tires and wheels on roads or rails. Some provision is made for sliding, but excessive sliding compromises the function of the surfaces. Normal operation involves little or no macroscopic slip. Static friction is often higher than the dynamic friction.

b. Clamped surfaces such as press-fitted pulleys on shafts, wedge-clamped pulleys on shafts, bolted joining surfaces in machines, automobiles, household appliances, hose clamps, etc. To prevent movement, high normal forces must be used, and the system is designed to impose a high but safe, normal (clamping) force. In some instances, pins, keys, surface steps, and other means are used to guarantee minimal motion. In the above examples, the application of a (friction) force frequently produces microscopic slip. Since contacting asperities are of varying heights on the original surfaces, contact pressures within clamped regions may vary. Thus, the local resistance to sliding varies and some asperities will slip when low values of friction force are applied. Slip may be referred to as micro-sliding, as distinguished from macro-sliding where all asperities are sliding at once. The result of oscillatory sliding of asperities is a wearing mechanism, sometimes referred to as fretting.

The works of all named authors in this chapter are described in reference 1 unless specifically cited.

2. *Energy absorption-controlling components* such as in brakes and clutches. Efficient design usually requires rejecting materials with low coefficient of friction because such materials require large values of normal force. Large coefficients of friction would be desirable except that suitably durable materials with high friction have not been found. Furthermore, high friction materials are more likely to cause vibration than are low friction materials. Thus, many braking and clutching materials have intermediate values of coefficient of friction, μ, in the range between 0.3 and 0.6. An important requirement of braking materials is constant friction, in order to prevent brake pulling and unexpected wheel lockup in vehicles. A secondary goal is to minimize the difference between the static and dynamic coefficient of friction for avoiding squeal or vibrations from brakes and clutches.
3. *Quality control components* that require constant friction. Two examples may be cited, but there are many more:

   a. In knitting and weaving of textile products, the tightness of weave must be controlled and reproducible to produce uniform fabric.
   b. Sheet-metal rolling mills require a well-controlled coefficient of friction in order to maintain uniformity of thickness, width, and surface finish of the sheet and, in some instances, minimize cracking of the edges of the sheet.

4. *Low friction components* that are expected to operate at maximum efficiency while a normal force is transmitted. Examples are gears in watches and other machines where limited driving power may be available or minimum power consumption is desired, bearings in motors, engines, and gyroscopes where minimum losses are desired, and precision guides in machinery in which high friction may produce distortion.

(See Problem Set question 6 a.)

EARLY PHENOMENOLOGICAL OBSERVATIONS[2]

Leonardo da Vinci (1452–1519), the man of many talents, also had some opinions on friction, specifically, $F \propto W$. After the start of the industrial revolution came the specialty of building and operating engines (steam engines, military catapults, etc.) and this was done by engineers. Amontons (1663–1705), a French architect turned engineer, gave the subject of friction its first great publicity in 1699 when he presented a paper on the subject to the French Academy. The science of mechanics had been under active development since Galileo (≈1600) and others. Amontons lamented the fact that "indeed among all those who have written on the subject of moving forces, there is probably not a single one who has given sufficient attention to the effect of friction in Machines." He then astounded his audience by reporting that in his research he found $F \approx W/3$ and F is independent of the size of the sliding body.

The specimens tested by Amontons were of copper, iron, lead, and wood in various combinations, and it is interesting to note that in each experiment the surfaces were coated with pork fat (suet). The laws enunciated by Amontons are frequently but inaccurately described by present day writers as the laws of "dry" friction and "it is a salutary lesson to find that the seventeenth century manuscript makes it clear that Amontons was in fact studying the frictional characteristics of greased surfaces under conditions which would now be described as boundary lubrication.[2]

EARLY THEORIES

Amontons saw the cause of friction as the collision of surface irregularities. The scale of these irregularities must have been macroscopic because little was known of microscopic irregularities at that time. Macroscopic irregularities were common and readily observed and in fact may be seen today on the surfaces of museum pieces fashioned in Amontons' day.

Euler (1707), a Swiss theologian, physicist, and physiologist who followed Bernoulli as professor of physics at St. Petersburg (formerly Leningrad), said friction was due to (hypothetical) surface ratchets. His conclusions are shown in Figure 6.1.

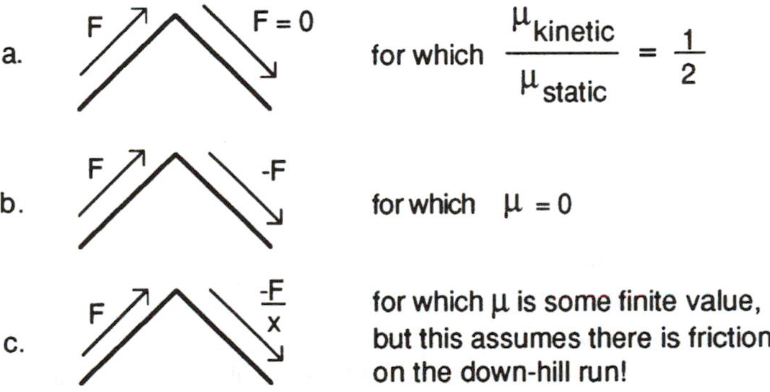

Figure 6.1 Sketch of Euler's description of friction.

Coulomb (1736–1806), a French physicist-engineer, said friction was due to the interlocking of asperities. He was well aware of attractive forces between surfaces because of the discussions of that time on gravitation and electrostatics. In fact, Coulomb measured electrostatic forces and found that they followed the inverse square law (force is inversely related to the square of distance of separation) that Newton had guessed (1686) applied to gravitation. However, he discounted adhesion (which he called cohesion) as a source of friction because friction is usually found to be independent of (apparent) area of contact. Again it is interesting to note that whereas Coulomb was in error in his explanation of friction, and he did not improve on the findings of Amontons, today "dry friction" is almost universally known as "Coulomb friction" in mechanics and physics. Perhaps it is well for this "error" to continue, for peace of mind. Without the prestige of Coulomb's name, the actual high variabilities of "dry" friction would be too unsettling. Coulomb and others considered the actual surfaces to be frictionless. This, of course, is disproven by the fact that one monolayer of gas drastically affects friction without affecting the geometry of the surfaces.

Samuel Vince, an Englishman (1749–1821), said $\mu_s = \mu_k$ + adhesion. An anonymous writer then asks whether motion destroys adhesion.

Leslie, also English (1766–1832), argued that adhesion can have no affect in a direction parallel to the surface since adhesion is a force perpendicular to the surface. Rather, friction must be due to the sinking of asperities.

Sir W. B. Hardy (works:1921–1928), a physical chemist, said that friction is due to molecular attraction operating across an interface. He came to this conclusion by experimentation. His primary work was to measure the size of molecules. He formed drops of fatty acid on the end of capillary tubes and measured the size of a drop just before it fell onto water. He then measured the area of the floating island of fatty acid on the water, from which he could determine the film thickness. One of these films was transferred to a glass plate. He found that the coefficient of friction of clean glass was about 0.6, but on glass covered with a single layer of fatty acid it was 0.06. He knew that the film of fatty acid was about 2 nm thick and the glass was much rougher. The film therefore did not significantly alter the functioning surface roughness but greatly reduced the friction. Hardy was also aware that molecular attraction operates over short distances and therefore differentiates between real area of contact and apparent area of contact.

Tomlinson elaborated on the molecular adhesion approach. The basis of his theory is the partial irreversibility of the bonding force between atoms, which can be shown on figures of the type of Figure 3.1 in Chapter 3.

In retrospect, friction research was accelerated with the publishing of an extensive work by Beare and Bowden. Their results were carefully checked with Tomlinson's and no correlation was seen. They proved that frictional effects are not confined to the first "molecular" layer and Tomlinson's work was dispatched with one statement: "It would appear that the physical processes occurring during sliding are too complicated to yield easily to a simple mathematical treatment." That may have been premature: there are several attempts under way to revive Tomlinson's approach.

DEVELOPMENT OF THE ADHESION THEORY OF FRICTION

Hardy's observation that one monolayer of lubricant reduces friction caused serious doubt about the validity of the idea that friction is due to the interlocking of asperities. The adhesion hypothesis was the best alternative in the 1930s although it was not clear which surface or substrate chemical species were prominent in the adhesion process. Several laboratories took up the task of finding the real cause of friction but none proceeded with the vigor and persistence of the Bowden school in Cambridge. The adhesion explanation of friction is most often attributed to Drs. Bowden and Tabor although there are conflicting claims to this honor. Usually the conflicting claims are supported by "proof" of prior publication of ideas or research results. On the other hand, it is easy to be mistaken in the presence of immature ideas and in the interpretation of research results, so full credit should not go to one who does not adequately convince others of his ideas. On the latter ground alone, Bowden and Tabor are worthy of the honor accorded them, Bowden for his prowess in acquiring funds for the laboratory and Tabor for the actual development of the concepts.

The adhesion theory was formulated in papers which were mostly treatises on the inadequacy of interlocking. Tabor advanced the idea that the force of friction is the product of the real area of contact and the shear strength of the bond in that region, i.e., $F = A_r S_s$. To complete the model, the load was thought to be borne by the tips of asperities, altogether comprising the same area of contact, multiplied by the average pressure of contact, $W = A_r P_f$. The average pressure of contact was thought to be that for fully developed plastic flow such as under a hardness test indenter, thus the subscript in P_f. Altogether,

$$\mu = \frac{F}{W} = \frac{A_r S_s}{A_r P_f} = \frac{S_s}{P_f} \qquad \textbf{(1)}$$

Both S_s and P_f are properties of materials. $P_f \approx 3Y$ and $S_s \approx Y/2$ and so the usual ratio S_s/P_f for ductile metals is between 0.17 and 0.2. A value of $\mu \approx 0.2$ is often found in practice for clean metals in air, but there are enough exceptions to this rule that Tabor's model came under considerable criticism. However, it was the first model that suggested the importance of the mechanical properties of the sliding bodies in friction.

Tabor then demonstrated the validity of the relationship $F = A_r S_s$ at least qualitatively by experiments with a hard steel sphere sliding over various flat surfaces as illustrated in Figure 6.2. Similar results have been found for wax on a hard surface, etc. This principle has been applied to the design of sleeve bearings such as those used in engines, electric motors, sliding electrical contacts, and many other applications. Engine bearings are often composed of lead-tin-copper-silver (and lately aluminum) combinations applied to a steel backing. The result is low friction, provided the film of soft metal has a thickness of the order of 10^{-3} or 10^{-4} mm, as shown in Figure 6.3

During the time of the development of the ideas on adhesion, the interlocking theory also had its supporters. The most vociferous was Dr. J. J. Bikerman who continued until his death in 1977 to hold the view that friction must be due to surface roughness. This view is based on the finding that sliding force is proportional to applied load. By itself this finding does not prove the interlocking theory. Bikerman agreed that the real area of contact should increase as load increases but insisted that it does not decrease as load decreases if there is adhesion. Thus, he would expect that friction would not decrease as load decreases if the adhesion theory is correct. Dr. Bikerman, an authority in his own right on the chemistry of adhesive bonding, had published his position as late as 1974 in the face of a continuous stream of evidence contrary to his conviction.[3]

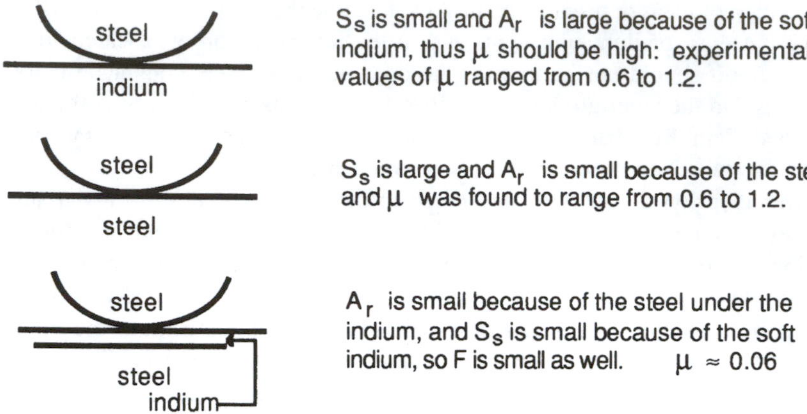
steel
indium
S_s is small and A_r is large because of the soft indium, thus μ should be high: experimental values of μ ranged from 0.6 to 1.2.
steel
steel
S_s is large and A_r is small because of the steel and μ was found to range from 0.6 to 1.2.
steel
steel
indium
A_r is small because of the steel under the indium, and S_s is small because of the soft indium, so F is small as well. μ ≈ 0.06

Figure 6.2 Demonstration of the F = AS concept.

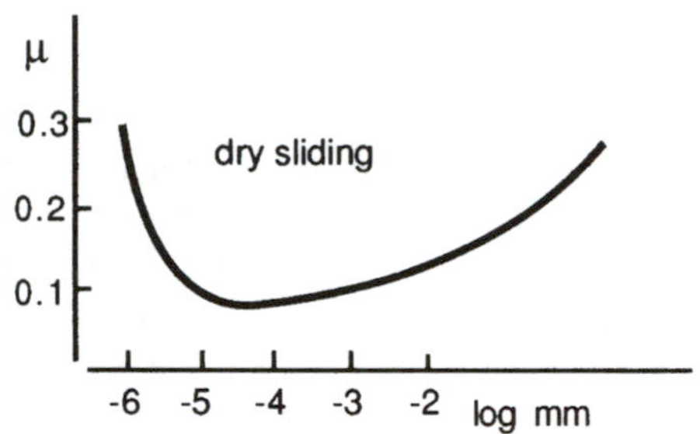
μ
0.3
0.2
0.1
dry sliding
-6 -5 -4 -3 -2 log mm

Figure 6.3 Influence of soft-film thickness on friction.

From 1939 to 1959 a series of papers appeared that provided the best arguments for the adhesion theory of friction. In essence, they show that for ductile metals, at least, asperities deform plastically, producing a growth in real area of contact which is limited by the shear stress that can be sustained in surface films. In effect, the coefficient of friction is determined by the extent to which contaminant films on the surface prevent complete seizure of two rubbing surfaces to each other. Bowden and Tabor showed, using electrical contact resistance, that plastic flow occurs in asperities even for small static loads. Bowden and Hughes further showed the role of surface species by measuring $\mu > 4$ in a vacuum of 10^{-6} Torr (0.133 mPa) on surfaces cleaned by abrasive cloth and by heating, whereas μ decreased considerably when O_2 was admitted to achieve a pressure of 10^{-3} Torr (0.133 Pa).

Further difficulties for the interlocking theory appeared in the findings of C. D. Strang and C. R. Lewis. Using large scale models they measured the energy required to lift a slider up to reduce interference of asperities and found that this requires only 10% of the total energy of sliding. E. Eisner measured the path of the center of mass of a slider as a pulling force increased from zero and found a significant downward displacement component, consistent with plastic flow of asperities. (See the discussion on plasticity in Chapter 2.)

The above findings led Rubenstein, Green, and Tabor to publish separate models for the plastic behavior of asperities. Tabor's is the most germane, however, and will be outlined below. The model begins with a two-dimensional asperity of non-work-hardening metal pressed against a rigid plate as shown in Figure 6.4. The initial load, W, is sufficient to produce plastic flow in the asperity, which produces a normal stress equal to the tensile yield strength, P_y, in the asperity, and a cross-sectional area of A_0.

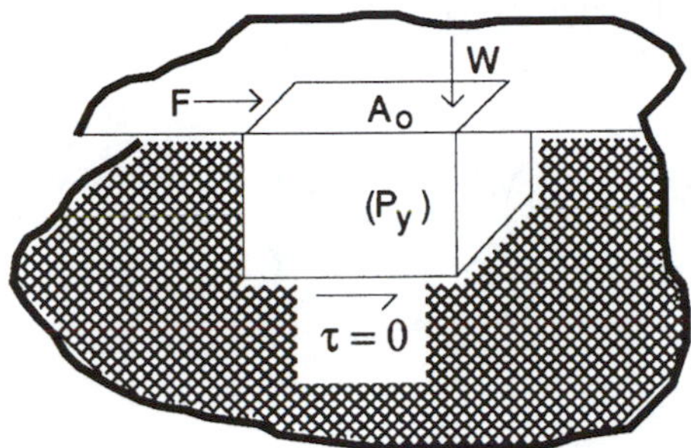

Figure 6.4 Tabor's model of a plastically deforming asperity.

At first the mean normal pressure is $P_y = W/A_o$, and F = 0 so that the shear stress, τ, is zero. Now apply a finite F (and the proper forces to prevent rotation of the element). Deformation does not respond to the simple addition of stresses in the element as if the material were elastic. Rather, deformation occurs in order to maintain the conditions for continued plastic flow. Tabor used the shear distortion energy flow criteria of von Mises in his work. By this theory, for the two-dimensional (plane strain) case, σ and τ are related by,

$$\sigma^2 + 3\tau^2 = K^2 \tag{2}$$

where K is comparable to the uniaxial yield strength of the metals. Initially $\tau = 0$, so $K = \sigma_y = P_y$. Because the material is already plastic, the addition of a very small τ will cause a decrease in σ via an increase in the area of contact from A_o to A. This continues so long as there is a tractive effort sufficient to increase τ.

For three-dimensional asperities in work-hardenable materials and for a non-homogeneous strain field (and contained plastic flow) the simple von Mises equations do not apply, but it can be expected that a relationship of the form

$$\sigma^2 + \alpha\tau^2 = K^2 \tag{3}$$

might be a good starting point. No exact theoretical solution for this case has yet come to light. However, approximations can be made. This model can be applied to real metals where the maximum value of τ is the shear strength S_s of the metal. The problem then is to find α. One method begins with $K \approx 5S_s$, the usually observed property of material. Then

$$\sigma^2 + \alpha\tau^2 = 25(S_s)^2 \quad \textbf{(4)}$$

For the specific case of very large junction growth, σ approaches 0 and τ approaches S_s; then for the general case $\sigma^2 + 25\,\tau^2 = K^2$ and $\alpha = 25$. But since this result is derived from measurements of P_y and S_s in plane stress, it doubtless does not apply directly to the actual complex stress state of Figure 6.4. Therefore, other means were sought to estimate α.

One approach is through experimental results. To do this the above equation was revised as follows:

$$\text{Set } \sigma = \left(\frac{W}{A}\right), \quad K = P_y = \left(\frac{W}{A_o}\right), \quad \tau = \left(\frac{F}{A}\right) \quad \text{so,} \quad \left(\frac{W}{A}\right)^2 + \alpha\left(\frac{F}{A}\right)^2 = \left(\frac{W}{A_o}\right)^2$$

which becomes

$$1 + \alpha\left(\frac{F}{W}\right)^2 = \left(\frac{A}{A_o}\right)^2 \quad \textbf{(5)}$$

Now define the general ratio, $\dfrac{F}{W} \equiv \Phi$

(not to be confused with $\left(\dfrac{F}{W}\right)$ for sliding, which we call μ)

and get

$$1 + \alpha\Phi^2 = \left(\frac{A}{A_o}\right)^2 \quad \textbf{(6)}$$

From experiments, one can find how much the contact junctions (regions) grow as F (i.e., Φ) increases but before sliding begins. This is shown in Figure 6.5.

To complete the analysis, Tabor estimated the values of α from various sources:

from work with the adhesion of indium	$\alpha \approx 3.3$
from work with electrical resistance of contacts	$\alpha \approx 12$
from the analysis above	$\alpha \approx 25$

Each value is suspect for good reason. Tabor selects $\alpha = 9$ because it has a simple square root, but it turns out that the conclusion reached from the analysis is more important than the actual value of α.

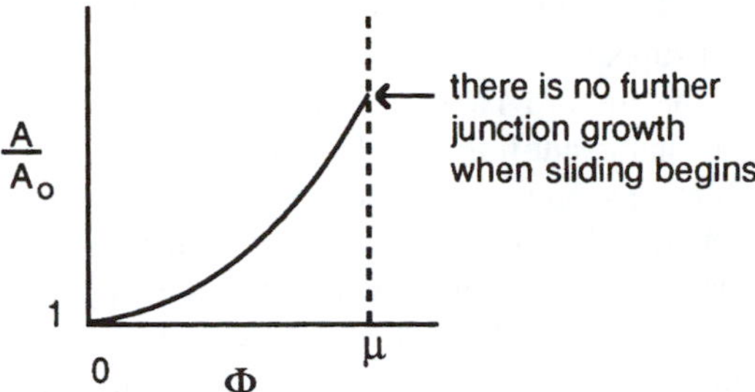

Figure 6.5 The manner by which junctions grow when F (i.e., Φ) increases.

Now assume that the surface contact region is weaker than the bulk shear strength, perhaps due to some contaminating film. Take the shear strength of the interface film to be S_i so that when the shear stress on the surface due to F equals S_i, sliding begins. Now since:

$$K^2 = P_y^{\ 2}$$

which = $\alpha(S_s)^2$, for the limiting case, using $\alpha = 9$,

$$\sigma^2 + 9S_i^{\ 2} = 9S_s^{\ 2}$$

Note that if S_i and σ operate over the same area of contact

$$\text{Taking}\left(\frac{S_i}{S_s}\right) = k, \text{ then}\left(\frac{\sigma^2}{S_i^2}\right) = 9(k^{-2} - 1), \text{ and } \left(\frac{S_i}{\sigma}\right) = \frac{1}{3\sqrt{k^{-2} - 1}}$$

and since both S_i and σ operate over the same area of contact:

$$\left(\frac{S_i}{\sigma}\right) = \mu = \frac{1}{3\sqrt{k^{-2} - 1}} \qquad \textbf{(7)}$$

Now we can see that if $k = 1$, $\mu = \infty$ which corresponds to clean surfaces, i.e., the junctions grow indefinitely and seizure occurs. But where

$k = 0.95$	$\mu = 1$
$k = 0.8$	$\mu = 0.45$
$k = 0.6$	$\mu = 0.25$
$k = 0.1$	$\mu = 0.03$

The study of the mechanisms of friction really becomes one of the study of the prevention of seizure! Or a study of the prevention of junction growth.

The equation $\mu = S_i/\sigma$ can be compared with the previous equation $\mu = S_s/P_y$. Not only is the ratio S_i/S_s likely to be less than one, but the ratio σ/P_y is as well.

In the new view $\sigma < P_y$ because the junction grows due to a shear stress. Recall that the new model supports the adhesion theory of friction mostly because the interlocking theory has no provision for plastic deformation of asperities or for the presence of a contaminant film with low shear strength.

Perhaps the ultimate support for the adhesion theory is embodied in the work of N. Gane.[4] By dragging the end of a fiber of tungsten over a surface of platinum, he was able to measure a friction force with a positive applied load, with a zero externally applied load, and finally with a negative applied load due to adhesion. His results press the definition of μ since he obtained values of positive μ, infinite μ, and negative μ, respectively, from these experiments.

(See Problem Set question 6 b.)

LIMITATIONS OF THE ADHESION THEORY OF FRICTION

The adhesion theory must be viewed as incomplete since to date it has not been useful for predicting real values of μ. In the model of Tabor, in Equation 7, it has not yet been possible to measure S_i except in a friction experiment, nor is the value of α known, as mentioned above. Even applying the expression F = AS to elastic materials misses the mark by at least a factor of 10, probably because the mode of junction fracture is not well understood.

The adhesion theory does not explain the effect of surface roughness in friction. The general impression in the technical world is that friction increases when surface roughness increases beyond about 100 micro-inches, although there are little reliable data to support this impression. Instantaneous variations in friction do increase in magnitude with rougher surfaces sliding at low speeds. The interlocking theory is not aided by the frequent observation that μ increases as surface finish decreases below 0.2 μm Ra. Bikerman explains this, however, by pointing out that the *fluid* film on all surfaces becomes important as a viscous substance on smooth surfaces.

The adhesion theory is so superior to the interlocking theory that it is easy to dismiss the influence of colliding asperities, particularly those composed of hard (second) phases in the micro structure. Several authors have published equations of the form:

$$\mu = \frac{S_s}{P_y} + \tan\theta \qquad \textbf{(8)}$$

The first term on the right is the same as that of Tabor, and θ is the average slope of plowing asperities. Derjaguin acknowledged the same effects in the equation F = μW + μAS where A is dependent on strain rate, temperature, etc. These then become two-term equations with a plowing term added to the adhesion term. Plowing was thought by some to cause up to one third the total friction force.

Another difficulty that the early adhesion theories of friction share with the classical laws of friction is that they apply to lightly loaded contact. Shaw, Ber, and Mamin show that for heavily loaded contact, such as in metal cutting, the friction stress may approach the simple shear strength of the substrate.[5] Apparently in heavily loaded contact $A_r \rightarrow A_a$, in which load-carrying asperities are closely spaced. The plastic field under each asperity is no longer supported by a large and isolated elastic field, which is the reason that $p_m \rightarrow 3Y$ in each contact region. The elastic fields under closely spaced asperities merge, or are coalesced, and in the limit become homogeneous as in a tensile specimen. Thus, $P_y \rightarrow Y$. Since in such cases $S \approx Y/2$, the highest value of $\mu \approx 1/2$. This assumption is widely used in metal working research. One consequence of this assumption is that $\mu = 0.5$ is often but erroneously considered to be the maximum possible value.

ADHESION IN FRICTION AND WEAR AND HOW IT FUNCTIONS

Is friction due to adhesion, or is it not? The question is far more important than a matter of favoring or rejecting the classic alternate explanation, namely the interference of asperities. The evidence that favors the adhesion explanation is actually rather direct, namely, that perfectly clean metals (in vacuum) stick together upon contact as discussed in Chapter 3.

The word "adhesion" is strongly embedded in the literature on friction and wear, probably because of such well-known equations as that of Tabor ($F = A_r S_s$) for friction and the equation of Archard ($\psi = kWV/H$) for wear rate. (See Equation 1, Chapter 8.) Adhesion is not often discussed as a cause of lubricated (viscous) friction though one could argue that wetting, surface tension, and even viscosity are manifestations of bonding forces as well.

Surely then, we are convinced that there is adhesion between any and every pair of contacting substances, though we do not know exactly how it functions. *All mechanisms of friction and wear should thus be referred to as adhesive mechanisms*. The fact that only a few are may mean that no other prominent cause or mechanism has been found for most cases.

It might be well to dispose of one argument concerning the word "adhesion". Coulomb, and later Bikerman, argued that friction could not be due to adhesion because adhesion is a resistance to vertical (normal) separation of surfaces, whereas friction is resistance to parallel motion of surfaces. Neither one denied that atomic bonding functions during sliding, but perhaps both should have coined a new term for this case.

ADHESION OF ATOMS

On the atomic scale, sliding is envisioned by some authors as the movement of hard-shell (and perhaps magnetic) atoms over each other as shown in Figure 6.6. Energy is required to move an atom from its rest position to the midpoint between two rest positions. However, that energy is restored when the atom falls into the next rest position. This cycle is thought to require no energy, and thus atom motion as shown cannot be the cause of friction.

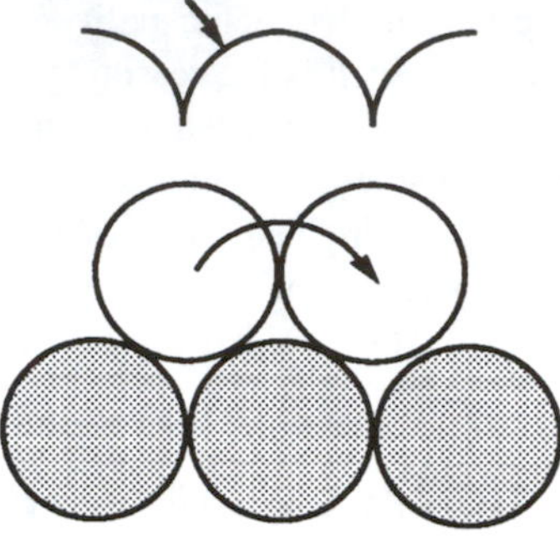

Figure 6.6 Magnetic ball model of sliding.

A more plausible explanation, for fairly brittle materials at least, involves atom A following atom B for some distance as atom B moves, as shown in Figure 6.7. This continues until the forces required to pull atom A, as atom B moves still further, exceeds that exerted upon atom A by its neighbors to keep it in position. At that point, atoms A and B separate. Atom A snaps back into position, setting its neighbors into vibration. Atom B snaps into the next rest position, setting its new neighbors into vibration. These lattice vibrations dissipate, heating the surrounding material, just as macroscopic vibration strains dissipate and heat a solid.

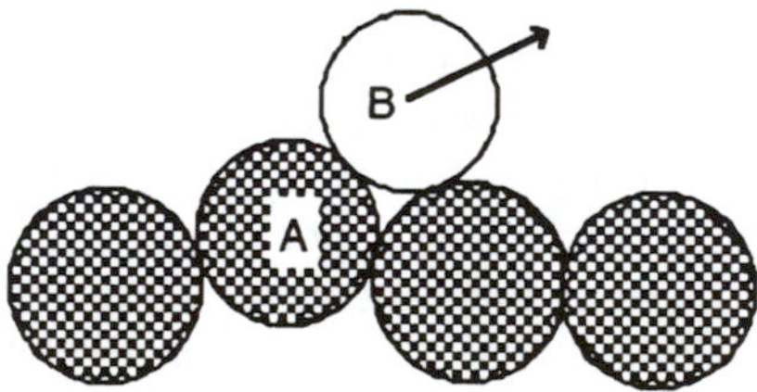

Figure 6.7 Movement of surface atom due to a slider.

In ductile materials atoms can be pulled even further out of position to produce slip, which, in macroscopic systems, is referred to as plastic flow. At this point it is helpful to make a comment for perspective. It would appear that ductile materials (metals, for example) would produce high friction, whereas brittle ceramic materials would produce low friction. In practice the opposite is usually found. These findings do not contradict the discussion of atomic friction: substances adsorbed upon solid surfaces of materials affect friction as strongly as do the substrate properties.

Friction also varies with direction of sliding on crystalline surfaces. In Figure 6.7 an atom moved from contact with two others, over the hump and back down into contact with two atoms again, all of them in the same plane. In a three-dimensional array of atoms, an atom is lodged in a well or pocket and in contact with three (or more) others. The single atom could move in many directions, locating wells at various spacings, requiring a significant range of energy exchange. This variation depends strongly on the bonding system for the material in question. There are four bonding systems: namely, the metallic bond, the ionic bond, the covalent bond, and the van der Waals bond systems. These bond systems are described in Chapter 3.

ELASTIC, PLASTIC, AND VISCO-ELASTIC EFFECTS IN FRICTION[1]

In discussions on the development of the adhesion theory of friction the emphasis was on the friction of those metals in which asperities become plastically deformed under even light average normal loads. The asperities of rubber, some plastics, wood, and some textiles appear to deform elastically. The consequence of the difference in behavior is as follows:

Plastically deformed asperities		Elastically deformed asperities
$A \propto W^1$ ←	Effect of load on (real) area of contact	→ $A \propto W^{2/3}$
$F \propto W^1$ ←	Adhesive friction force	→ $F \propto W^{2/3}$
μ = const. ←	Coefficient of friction	→ $\mu \propto W^{-1/3}$

The above are idealized cases to some extent. For a soft metal covered by a brittle oxide it has been found that there are three regimes of friction over a range of load. In Figure 6.8, in regime A the oxide film is intact, in regime C the oxide film is fractured, and regime B is a transition region.

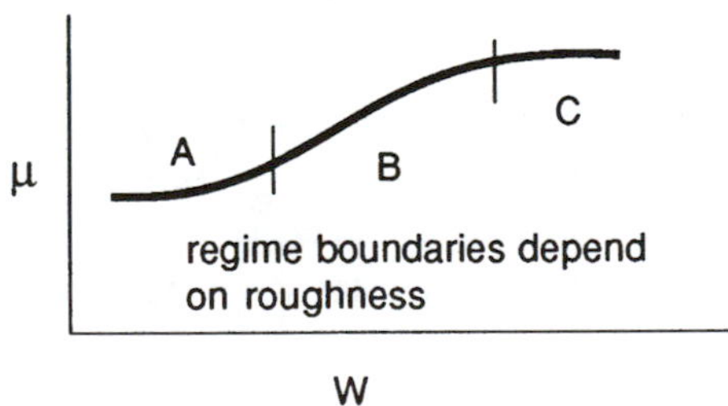

Figure 6.8 The influence of applied load on friction, for metals with brittle oxides.

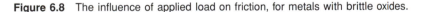

Visco-elastic materials such as rubber and plastics, show interesting friction properties that may vary by a factor of 5 to 1, or even 10 to 1 over a range of sliding speed or over a range of temperature. For example, Grosch slid four types of rubber on glass, yielding results of the type sketched in Figure 6.9.[6] When these data are transformed by an equation known as the WLF equation (see *Visco-elasticity* in Chapter 2) one master curve is formed as shown in Figure 6.10. This master curve has the same half-width as the visco-elastic loss peak for the same rubber, which suggests that the same phenomenon is operating in sliding friction as in material irreversibility (hysteresis loss) in a vibratory test. Grosch shows that 1 cm/sec sliding speed is equivalent to 6×10^6 c.p.s. of vibration; and he takes this to mean that the surfaces of sliding rubber are jumping along rather than sliding. This implies a surprisingly narrow spectrum of vibrations which seems unlikely. The vibrations of Grosch may correspond with the waves of detachment described by Schallamach[7] and discussed later in this chapter. By the model of Schallamach there need be no actual sliding of rubber over glass to effect relative motion. Rather, the rubber progresses in the manner of an earthworm, and the coefficient of friction may be due to damping loss in the rubber and irreversibility of adhesion.

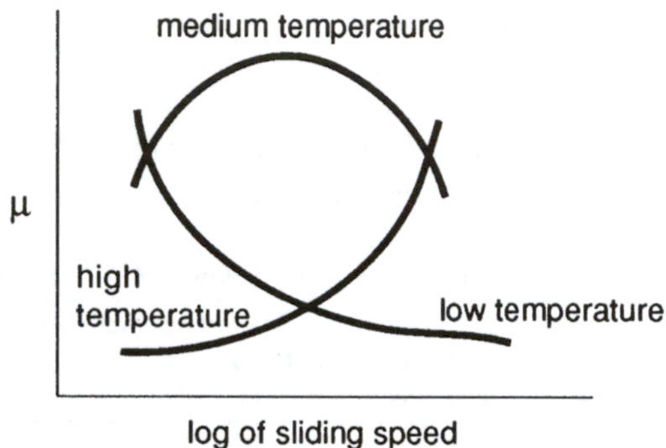

Figure 6.9 Friction of rubber on glass in three temperature ranges. (Adapted from Grosch, K.A., *Proc. Roy. Soc.*, A274, 21, 1963.)

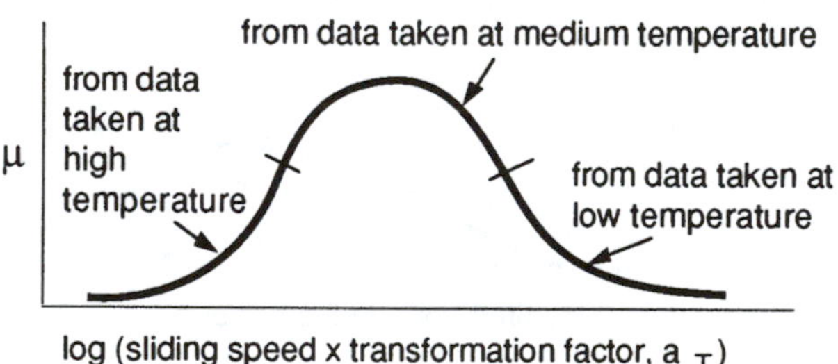

Figure 6.10 Data from Figure 6.9 transformed by visco-elastic transforms.

Most theories of the friction of polymers are based on continuous contact of sliding surfaces. However, some are based on concepts derived from chemical kinetics. Schallamach explains rubber friction as being due to "activation processes." He found that friction curves transform along the sliding-speed axis in response to temperature change according to the Arrhenius equation $V = V_o e^{-Q/RT}$ for rubber. (The Arrhenius equation is useful but not precise over a very wide range of temperature. The WLF equation is better only between T_g and $T_g + 100°C$.) Each release of bond and formation of a new one is conditioned by an activation process.

Results almost identical to those of Grosch were measured for acrylonitrile-butadiene rubber.[8] *If* $F = A_r S_s$, the variation in friction with temperature and sliding speed must reflect the variation in A_r and S_s with strain rate and temperature. Data are available from which A_r and S_s can be inferred. Data for the fracture strength of a styrene-butadiene rubber are sketched in Figure 6.11.

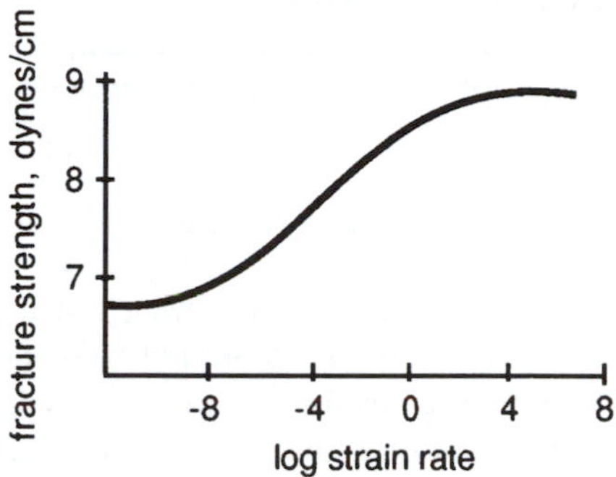

Figure 6.11 Fracture strength versus strain rate for rubber.

This curve is also transformable by the WLF equation. Now A can be estimated by remembering that $a \propto 1/E^{1/3}$ so $A_r \propto 1/E^{2/3}$. Data for E for the same rubber are given with a corresponding curve for A_r in Figure 6.12. A_r and S_s can be multiplied graphically to get F. But this produces a fairly straight line, as shown in Figure 6.13a, if the transitions in A_r and in S_s are coincident on the strain rate axis.

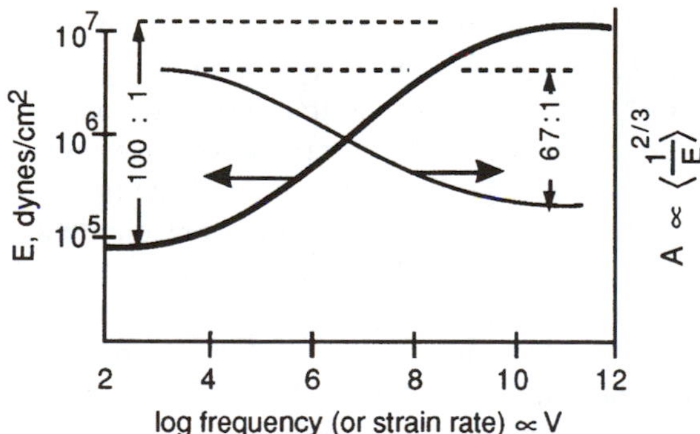

Figure 6.12 Variation in elastic modulus over a wide range of vibration frequency.

A different conclusion can be reached, however, based on the mechanics of the friction process. The variation in A_r is controlled by the strain rate relatively deep in the substrate. The rate of strain in the substrate is therefore some low multiple of the sliding speed, whereas the rate of strain in the asperities must be some high multiple of the sliding speed. Thus for a particular sliding speed, the strain rate in the shearing layer at the interface is high and the strain rate in the

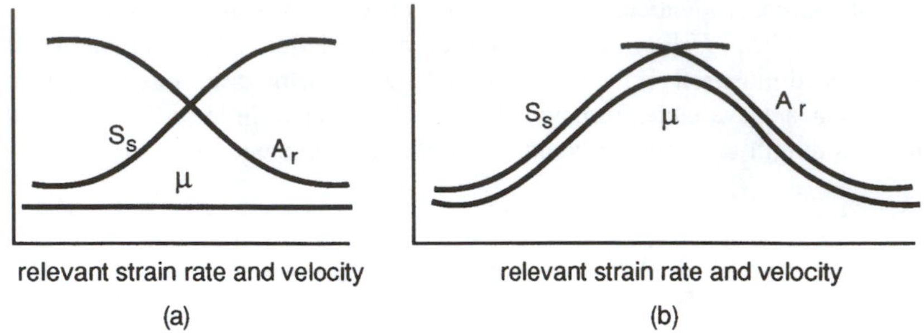

Figure 6.13 Showing the influence of displacement of curves S_s and A_r.

substrate, which controls the value of E, is lower. For a given sliding speed, therefore, the transitions in the two curves are not coincident. The curve for S_s reaches a high value of S_s at a relatively low sliding speed, i.e., the curve for S_s should be shifted to the left relative to the curve for A_r. A fair estimate is that the shear rate in the surface layer would be 5 to 6 orders of 10 higher than the average shear strain rate in the substrate when a slider slides. This is shown in Figure 6.13b where a curve for μ or F produces a peak, and when worked out precisely, the peak has eight times the magnitude of the background. This would support the suggested mechanics of friction.

Several experimental observations in the sliding of rubber are not yet explained. For example, it is sometimes observed that the coefficient of friction changes after a speed change, but not immediately. Schallamach calls this effect "conditioning."

The above variations in rubber friction are usually satisfying because of the large effects seen in experiment. Interesting effects are also seen in the linear polymers (or plastics) below T_g. Above T_g most linear polymers are viscous liquids, and below T_g there are structural transitions not found in rubber, which requires some caution. The friction data for plastics often show rather mild slopes and often only suggestions of peaks, even when the experimental variables cover a very wide range. The curves do not transform as readily to a master curve as was shown above with rubber. In addition, as found by Bahadur,[9] morphological changes that occur in the polymer due to temperature change necessitate a vertical shift in data curves in addition to the horizontal WLF type of shift to produce a master curve. Nonetheless, the data for several polymers are interesting to study. The most notable points are that the coefficients of friction do indeed vary considerably for linear polymers and that only in rare instances do the measured coefficients of friction compare with those given in handbooks. For example, Figures 6.14 and 6.15 show the coefficient of friction for a wide range of sliding speed (below 1 cm/sec to avoid frictional heating) and test temperature for PTFE, polyethylene, and Nylon 6-6. The handbook value for the coefficient of friction for PTFE is 0.07, and for the others is 0.39.

The more rigid thermo-setting polymers show no interesting variations in friction at the low speeds (<1 cm/sec) used in experiments with rubber and linear polymers. Some work has been done with thermo-setting resins at higher speeds

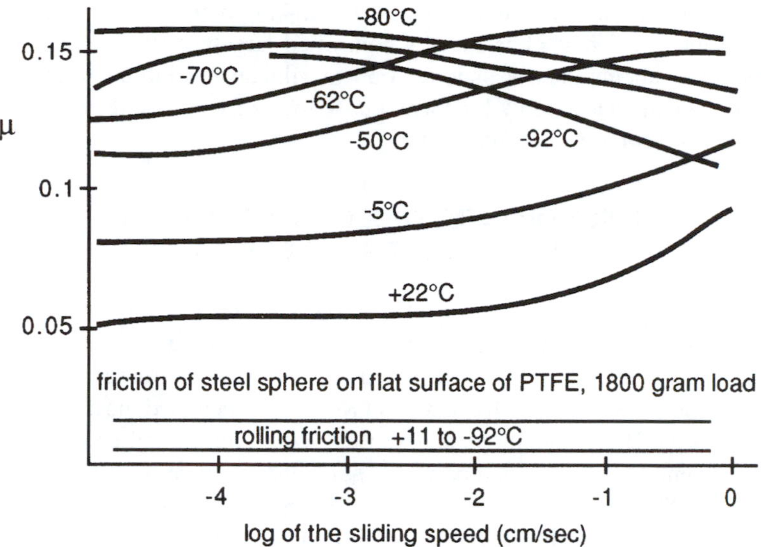

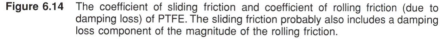

Figure 6.14 The coefficient of sliding friction and coefficient of rolling friction (due to damping loss) of PTFE. The sliding friction probably also includes a damping loss component of the magnitude of the rolling friction.

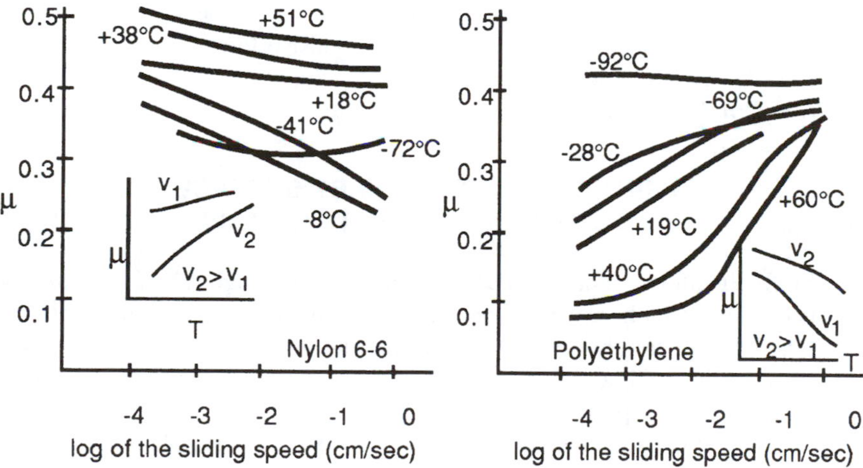

Figure 6.15 The coefficient of sliding friction for Nylon 6-6 and polyethylene.

usually associated with the speed at which automotive brakes operate. Thermosetting polymer is one of the several constituents in brake materials, and is often the binder for asbestos, metal chips, Kevlar fiber, and other additives. For safe and comfortable operation of vehicles it is necessary that the coefficient of friction of brake materials be constant in each wheel, with time and over a production lot. In addition, it is necessary that the coefficient of friction does not increase as temperature increases to prevent wheel lock-up, and it should not decrease (fade) at the high temperatures achieved by braking down a long hill. In brake material the coefficient of friction is controlled largely by the nature of wear debris in the rubbing interface and the transfer film attached to the rotating metal member, which considerably broadens the scope of friction studies.

(See Problem Set question 6 c.)

FRICTION INFLUENCED BY ATTRACTIVE FORCES BETWEEN BODIES

In careful work on the area of contact between soft smooth rubber and smooth glass, the area of contact was found to be larger than could be accounted for by the Hertz calculation. This was attributed to van der Waals forces attracting the rubber to the glass. Johnson, Kendall, and Roberts[10] calculated the area of contact using both the Hertz conditions and van der Waals forces and came very close to experimental observations. For a very soft rubber sphere of effective Young's Modulus of 8×10^6 dynes/cm^2 (812 KPa) and applied load of 500 grams, the van der Waals forces add 45 grams to the load, thus increasing the coefficient of friction by 9% over that defined by $\mu = F/W$. This effect would probably diminish by one order of ten for every decade of increase in Young's Modulus and be negligible for such substances as tire rubber.

An opposite effect may be seen when rubber slides on glass that has been wetted by water containing, for example, Na^+ and Cl^- ions. Each solid may attract the same polarity ions, which produces a net repulsive force, reducing the measurable coefficient of friction.[11]

(See Problem Set question 6 d.)

FRICTION CONTROLLED BY SURFACE MELTING AND OTHER THIN FILMS

Surface melting might be expected to occur at very high rubbing speeds and in such cases the molten material on the surface could be considered a lubricant. Such melting apparently occurs between the ring on the bourtolet of shells and the barrels of big military guns.[12] These rings, formerly of gilding metal (brass) and more recently of polymers, are single purpose components, simply to engage in the rifling of the gun tube, so their wear is of little concern. Melting doubtless occurs on the surface of polymers more readily than on metal surfaces because metals have much higher thermal conductivity than do polymers.

A widely known case of melting at the sliding interface is that between skates and ice. Ice is actually a rather complex visco-elastic substance. Data shown in Figure 6.16, for the friction of steel on ice show surprisingly high values at low temperatures.[13]

Ice is covered with a water layer above –25°C which becomes thicker as temperature rises. This water has the O_2 preferentially oriented outward. Ice is ordinarily of hexagonal structure with a high surface energy. Some reorientation occurs on the surface to lower surface energy to the extent of changing lattice

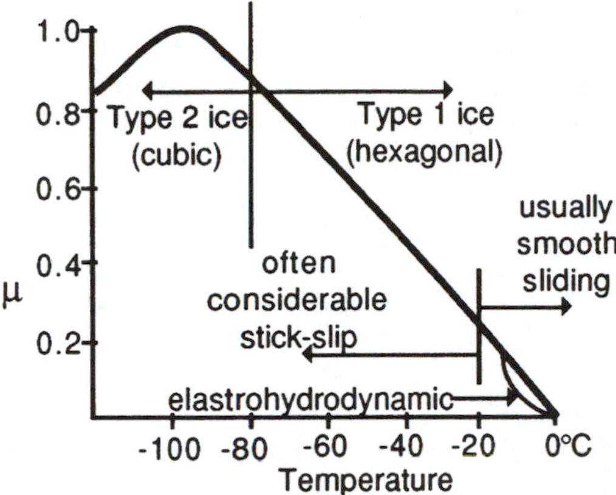

Figure 6.16 The coefficient of friction of a steel pin on ice over a temperature range.

form, and there is still sufficient energy to orient the water film. Since melting of ice to water involves a reduction in volume, a slider which applies a normal stress encourages surface melting.

Adsorbed gas, water vapor layers, and organic contaminant films surely influence friction. Their effects could be considered those of lubrication, though to formalize concepts in this topic it would be necessary to characterize the thin films in terms of their thickness and viscosities. Friction is often seen to vary with humidity and is influenced by such factors as, the amount of handling of specimens with human hands, cleaning methods, method of storage, number of passages of the slider, and many other factors. The films on such surfaces vary in thickness up to 300 nm, but are invisible and thought not to be important.

In one extreme example of the influence of surface films, and perhaps other factors as well, a polyurethane of 70 Shore A hardness was pressed against the flat surface of a 12-inch-diameter quartz disc (in air), which was turned at about 5 RPM. In one 90° segment of the disc μ was 4, in the second quadrant μ was 10, in the third segment μ was 4, and in the fourth, μ was 10 again. This experiment was observed by several seasoned research engineers and physicists with great wonder! Upon reversing the direction of rotation of the disc the pattern was repeated but shifted backward (relative to the first turning direction) by 45°.

ROLLING RESISTANCE OR ROLLING FRICTION[1]

Rolling resistance arises from two sources, sliding of one contacting surface along the other, and irreversibility in the deformation of contacting materials.

Rolling of a sphere or cylinder along a flat surface can be viewed as a series of indentations along the flat surface. When a steel sphere indents a slab of rubber, the rubber stretches in the indented region but the steel does not. Thus there is sliding between the steel and rubber. Reynolds pointed this out for the case of *spheres and cylinders.*

Heathcote noted a second mode of sliding between a *sphere* and a slab of rubber. In this case the sphere advances a distance less than πD (D = the diameter of the sphere). The reason is that the instant center of rolling is just above the lowest point of contact as shown in Figure 6.17, and there is sliding in the area of contact.

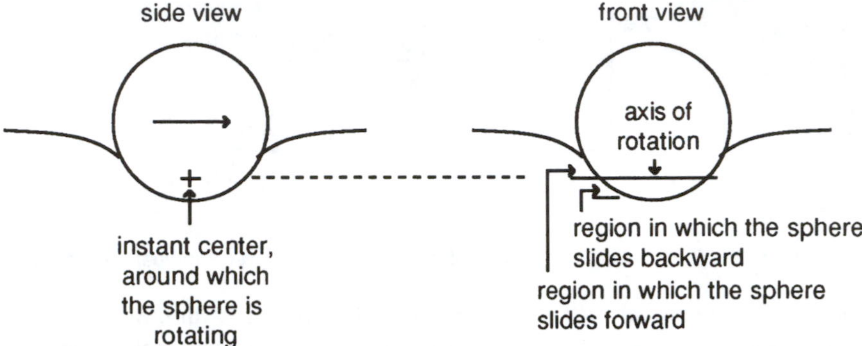

Figure 6.17 Mechanics of the rolling of a sphere on a soft flat plate. Above the instant center, or axis of rotation, the sphere slides forward, and below the instant center the sphere slides backward.

These two types of slip were, at one time, considered to be the chief causes of rolling friction. However, this is not supported by experiments that show lubricated rolling friction to be very nearly the same as dry rolling friction. Neither is it supported by experiments with the two (or four) ball pendulum tester (see Figure 6.18) where neither Heathcote slip nor Reynolds slip are present. (If the balls have the same γ, E, and R, the contact between them is planar.)

Figure 6.18 The pendulum test for measuring the damping loss of material.

In such tests and in lubricated rolling of a roller over a flat surface a significant and widely varying rolling friction may be seen. This is due to either elastic hysteresis, visco-elasticity, or plastic deformation.

Tabor examined the strain state under a roller for the elastic case and concluded that an element of material in the substrate of the flat body passes through 3.5 strain cycles as a roller moves along the surface. This idea is shown in Figure 6.19.

Rubber was also cycled in tension and release to measure the fraction of strain energy which was put into the tensile specimen during loading but not recovered

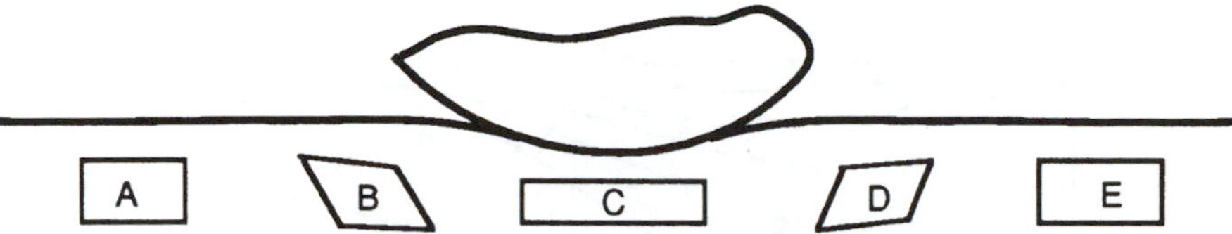

from A to C there is 1 cycle of shear and release
from C to E there is 1 cycle of shear and release
from A to E there is 1 cycle of compression and release
from A to C to E there is 1/2 cycle (45°) of shear rotation and release, making a total of 3.5 cycles.

Figure 6.19 Strain cycles in a flat plate over which a cylinder rolls.

upon release of load. The fraction of energy lost was designated as α. Tabor calculated the expected rolling resistance, F, as the energy required for the front half of the cylindrical roller to push rubber down to the maximum depth of roller indentation multiplied by the fraction of energy lost per cycle of strain in rubber, α, and multiplied by the number of cycles of strain experienced by the rubber in the substrate, β.

Thus

$$F = \alpha\beta\left(\frac{2Wa}{3\pi R}\right) \quad \textbf{(9)}$$

Rollers of a different shape than a long cylinder would be expected to produce more complicated strain fields in the rubber, not readily quantified. Experiments were done with four rollers of different shapes, and the values found for β were as follows:

Roller	β
Long cylinder	3.3
Short cylinder	2.9
British penny	2.0
Sphere	2.2

The smaller values for β for the short bodies were probably due to the rubber moving laterally from under the roller to avoid severe straining.

Tabor used α as a fixed quantity, which is not valid for situations where strain rates or temperature vary over a wide range. Damping loss (variously given in terms of tan δ or Δ, in distinction to α) varies with strain amplitude as well as strain rate and temperature. A typical plot of the effect of strain rate and strain amplitude is shown in Figure 6.20. In the rubber substrate under a roller there is a wide range of strain amplitude and strain rate, so that a strictly analytical calculation of rolling loss would be very complicated.

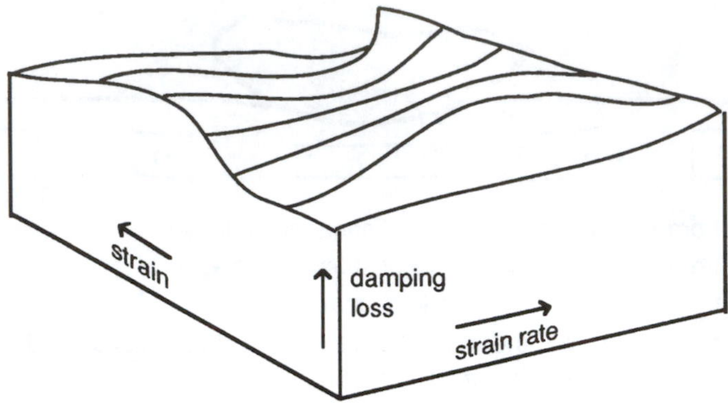

Figure 6.20 Sketch of the influence of strain rate and strain (amount) on damping loss of rubber.

In the case of rolling a metal roller on a visco-elastic material, there is a prominent effect due to rolling speed as shown in Figure 6.21. The shallow depth of indentation at high speed reflects the high effective elastic modulus at the strain rate in the flat substrate and vice versa for the low speed. The slow recovery of the flat plate material at medium speed reflects the higher damping loss in the material at intermediate strain rate than at high or low strain rates.

Figure 6.21 Rolling on a visco-elastic material, at three different speeds.

The case of rolling where there is plastic flow in the flat surface differs from the elastic case, as shown in Figure 6.22. In the elastic case there would be no evidence of the indentation after the roller has passed, except in the case of a great many cycles of rolling there could be some fatigue damage.

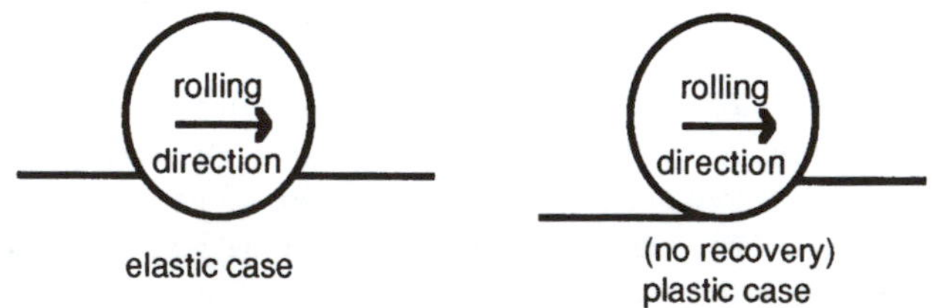

Figure 6.22 Rolling on an elastic material leaves no permanent indentation.

FRICTION OF COMPLIANT MATERIALS AND STRUCTURES, AND OF PNEUMATIC TIRES

The start of sliding is complicated by the elastic distortion of one or both of the sliding objects. With stiff systems and with imprecise instruments it appears as if sliding begins as a step function from no sliding when no force is applied, to complete sliding when a force is applied. Actually, as a prime mover begins moving, the resisting force increases with displacement, until sliding *seems* to occur.

The progression toward complete sliding is particularly important in hydraulic cylinders, for example, which have rubber seals (O-rings or other shapes) between the moving member and the stationary member. When an input (for example, volume and pressure of hydraulic fluid) must be controlled very precisely in order to achieve some desired output, seal compliance and friction start-up should be well characterized. Compliance can be estimated from the mechanical properties of the seal, generally, but the friction behavior cannot yet be predicted.

The same behavior may be seen in materials of high Young's Modulus, though the system compliance is usually too small to be observed. Actually, sliding occurs progressively over most contacting surfaces, rather than instantaneously over the entire contact area, unless they are very carefully made to avoid this effect. The progressive nature of sliding is often seen when a reciprocating force is applied that is less than sufficient to cause complete sliding: the center of contact will be dull in appearance, whereas the surrounding region will be shiny.

To illustrate, press two steel spheres together at $P_m < 1.1$ Y (end of the elastic range). The asperities will deform plastically but the overall (global) deformation will be elastic. The contact stress distribution will be elliptical. Now, when a force F is applied, according to Mindlin, a uniform shear stress is applied over the contact area except at the edges. However, the shear stress (traction) cannot exceed μP. This condition is shown in Figure 6.23. Outside of a central region there will be slip. Slip occurs in the nonshaded regions shown in Figure 6.24.

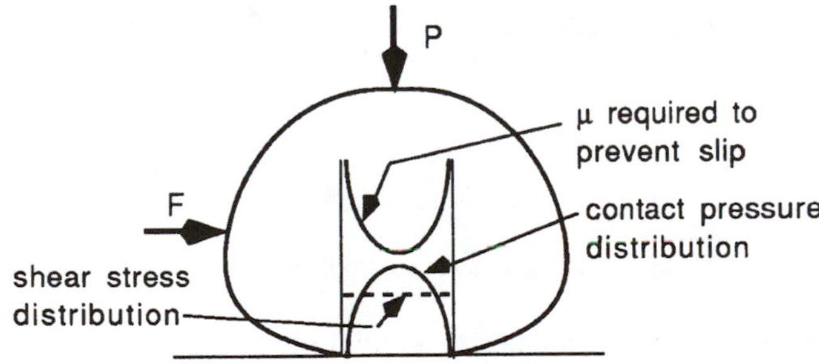

Figure 6.23 Stress distribution with normal and friction forces applied.

1. *Pneumatic tires* are compliant structures and the contact pressure against the road surface is nonuniform. The contact pressure distribution for a standing tire

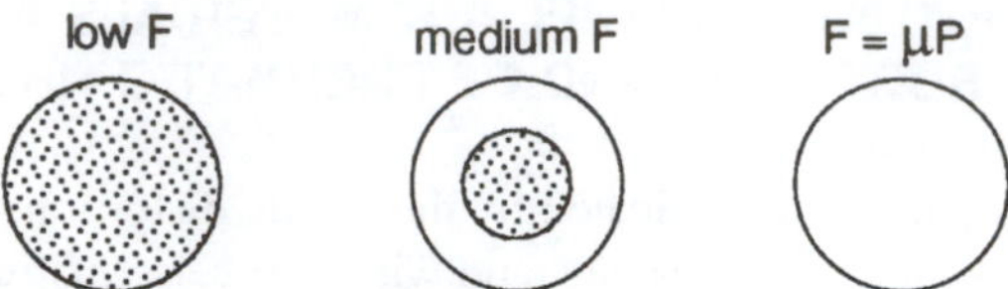

Figure 6.24 Region of no-slip (shaded) and slip (clear) when various forces are applied.

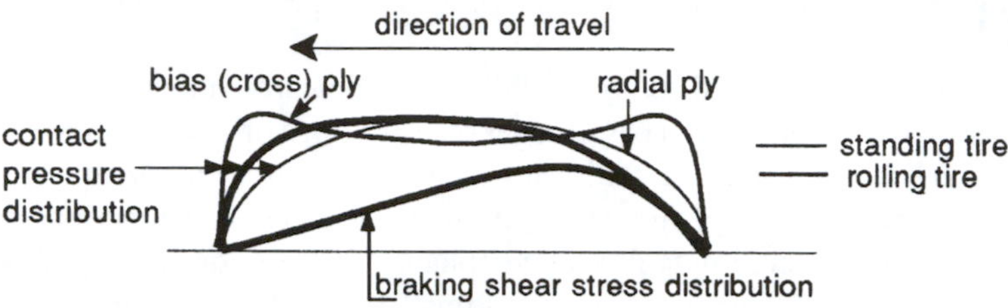

Figure 6.25 Longitudinal contact pressure distribution between a pneumatic tire and road surface, and the friction shear distribution for a braked tire.

is shown schematically in Figure 6.25 for two prominent types, the crossply and the radial ply tires. The general shape of pressure distribution is about the same in both the longitudinal and lateral directions.

A freely rolling tire has a skewed contact pressure distribution, as shown also in Figure 6.25, for the radial ply tire. Some rolling resistance comes from the visco-elastic damping loss in the tire carcass mostly, with the functional effect of moving the center of pressure ahead of the axle of the wheel as shown in Figure 6.26.

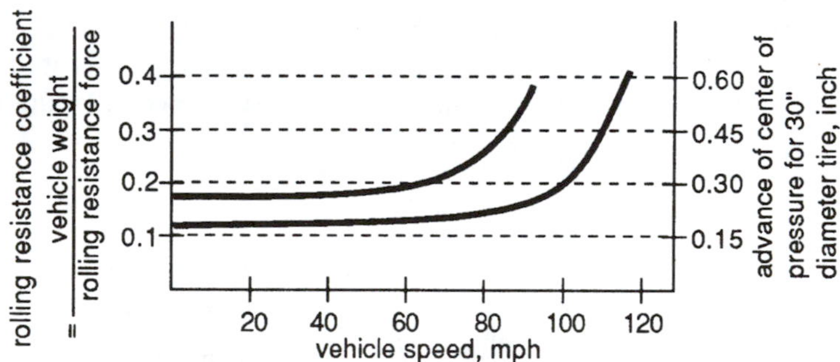

Figure 6.26 Range of rolling loss for automotive tires over a range of speed.

When a friction force is applied to a radial ply tire (as in braking or accelerating), slip occurs around the outer zone of contact with the road surface, but not symmetrically because of the nonuniform pressure distribution, and also because braking distorts the sidewalls of the tire so that the contact patch is pulled toward the rear of the axle. The friction forces (traction stresses) increase from the front of contact toward the rear of contact, also as shown in Figure 6.25. With increased application of braking torque, sliding or slip begins at, and grows from, the rear of contact.

It would appear from the friction force curve in Figure 6.25 that the overall braking force would increase as the region of slip or sliding grows toward the front of contact. Actually the point of maximum braking force occurs in the range from about 10% to 20% slip, depending on the type of tire, the load on the tire, the inflation pressure, and the skid resistance number of the road surface. This behavior is sketched in Figure 6.27, and it has not yet been satisfactorily explained.

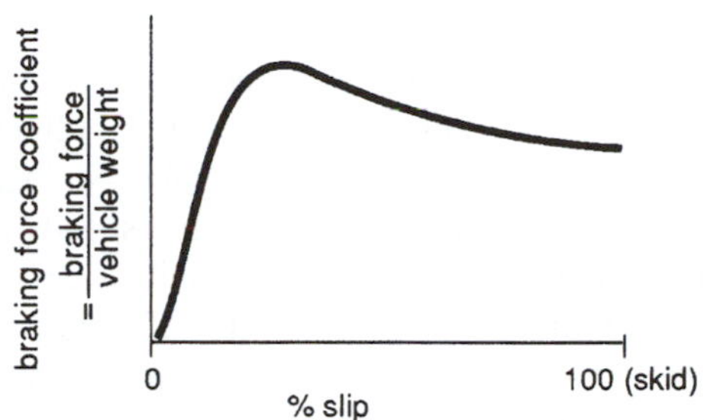

Figure 6.27 Braking force versus % slip for automotive tires.

Slip may be defined in terms of rotational speed. For example, for a vehicle moving at 60 mph, the wheels will rotate at a rate equivalent to a vehicle speed of 40 mph for 33% slip. Specific values of friction force capabilities of tires on dry and wet roads will be given in the Section titled *Tire Traction on Wet Roads* in Chapter 7.

2. *Discontinuous nature of sliding of some elastomers.* Slip or sliding of elastomers along hard surfaces sometimes proceeds by a very interesting mechanism. The most revealing experiments were done with a very soft rubber sphere against glass. In static contact the contact area is circular. When lateral (friction) force is applied the contact area diminishes, mostly by separation of rubber at the front of the contact region, where the rubber is in tension. A small amount of slip then occurs in a uniform ring around the center, except for the separated region at the front. The rubber at the rear of contact is in compression and it buckles, under the proper conditions, much as a rug buckles when pushed along the floor. The buckle moves from the rear of contact to the front, and has the effect of moving rubber along the glass without actually sliding along the glass. These buckles are referred to as Schallamach waves. Several waves may cross the contact area at the same time, and each one constitutes a moving strain field and nonuniform motion. When enough energy is developed in these waves, a sound can be heard. This is probably the source of the squealing of tires (and of sport shoes on smooth floors).

3. *Conclusions.* The general conclusion available from the observations described in this section is that friction is clearly not adequately described by a coefficient. Neither should any informed person force Coulomb friction into an analysis

unless it is decided beforehand that a good solution is not required for the problem at hand. System modelers are particularly vulnerable in this area. It appears that, out of long habit, models of mechanical systems provide space for only a single value of friction. Model makers become rather desperate to find that "right" value, wherever it may be found.

(See Problem Set question 6 e.)

THE INFLUENCE OF SOME VARIABLES ON GENERAL FRICTIONAL BEHAVIOR

Almost all operating parameters (speed, load, etc.) will influence the coefficient of friction. Some of the variables and their general effects are listed below.
1. *Sliding speed.* For metals and other crystalline solids sliding on like materials, the behavior is as shown in Figure 6.28. The sliding speeds indicated in Figure 6.28 range from imperceptibly slow (the tip of the minute hand on a watch moves at about 10^{-3} cm/sec) to normal walking speed (~125 cm/sec or 250 f/m) which covers many practical conditions. At very high-sliding speed (>2500 cm/sec) surface melting may occur to produce a very low coefficient of friction.

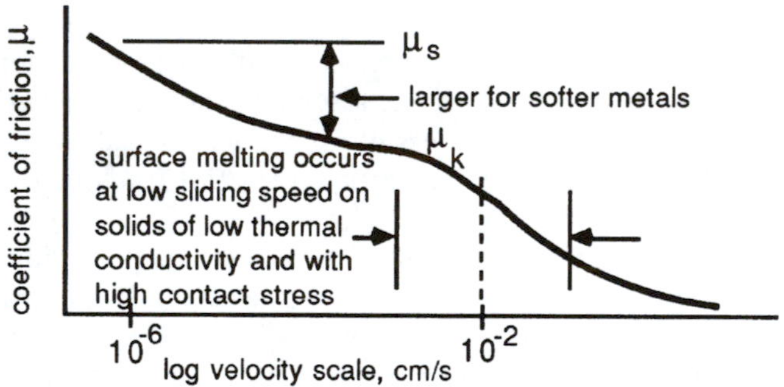

Figure 6.28 Frequently observed reduction of friction with sliding speed for crystalline solids.

Some polymers behave as shown in Figure 6.29 which is for the coefficient of friction of a steel sphere sliding on PTFE and Nylon 6-6. Note the variation for PTFE, which is usually thought to have a low and constant coefficient of friction. The coefficient of friction of both polymers increases with sliding speed over a limited range of speed because sliding evokes a visco-elastic response from the materials.
2. *Temperature.* There is usually little effect on the coefficient of friction of metals until the temperature becomes high enough to increase the oxidation rate (which usually changes μ). Increased temperature will lower the sliding speed at which surface melting occurs (see Figure 6.28) and increased temperature will shift the curve of coefficient of friction versus sliding speed to a higher sliding speed in many plastics (see Figures 6.9, 6.14, and 6.15).

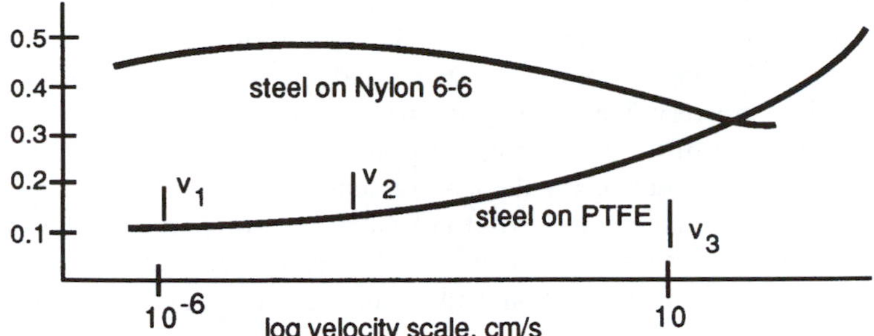

Figure 6.29 Master curves for the coefficient of friction of Nylon 6-6 and PTFE, values taken from Figures 6.14 and 6.15.

3. *Starting rate.* Rapid starting from standstill is sometimes reported to produce a low initial coefficient of friction. In many instances, the real coefficient of friction may be obscured by dynamic effects of the system holding the sliding member.

4. *Applied load or contact pressure.* In the few instances in which the coefficient of friction is reported over a large range of applied load, three principles may be seen in Figure 6.30. The first is that the coefficient of friction normally decreases as the applied load increases. For clean surfaces, as shown by curve "a," values of μ in excess of 2 are reported at low load, decreasing to about 0.5 at high loads. As mentioned earlier, in theory at least, very high average contact pressure should produce $\mu \approx 1/2$. Practical surfaces, as represented by curve "b," usually have values less than 1/2 because of surface contaminants. If the surface species include a brittle oxide, chipping off the oxide can expose clean substrate surfaces which increases local adhesion to cause higher coefficients of friction as shown in curve "c." It should be noted that some oxides are ductile under the compressive stresses in the contact region between hard metals. If these oxides are soft they may act as lubricants. If they are hard they may inhibit sliding. For example, a commercial black oxide on steel in a press fit increases dry friction by 50% or more.

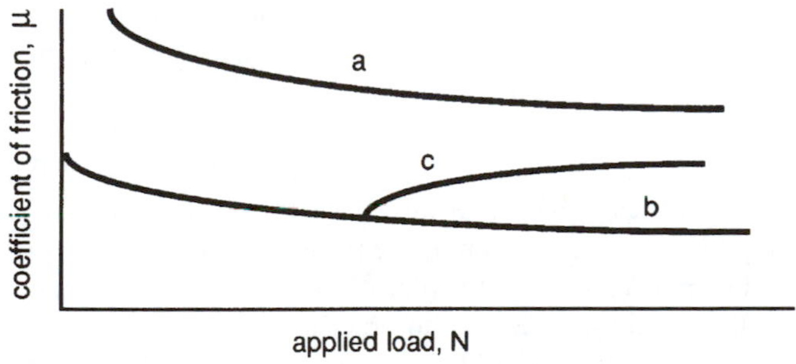

Figure 6.30 Three common influences of contact pressure on friction.

5. *Surface roughness* usually has little or no consistent effect on the coefficient of friction of clean, dry surfaces. Rough surfaces usually produce higher coefficients of friction in lubricated systems, particularly with soft metals where lubricant films are very thin as compared with asperity height.
6. *Wear rate*. One of the few consistent examples relating high coefficient of friction with surface damage is the case of scuffing. Galling and scoring also produce a high coefficient of friction usually accompanied by a *severe* rearrangement of surface material with little loss of material. In most other sliding pairs there is no connection between the coefficient of friction and wear rate.

STATIC AND KINETIC FRICTION

The force required to begin sliding is often greater than the force required to sustain sliding. One important exception is the case of a hard sphere sliding on some plastics. For example, for a sphere of steel sliding on Nylon 6-6, μ at 60°C varies with sliding speed as shown in Figure 6.29. The "static" coefficient of friction is lower than that at v_2. Most observers would, however, measure the value of μ at v_2 as the static value of μ. The reason is that v_1 in the present example is imperceptibly slow. The coefficient of friction at the start of visible sliding at v_2 is higher than at v_3. In this case it may be useful to define the *starting* coefficient of friction as that at v_2 and the *static* coefficient of friction as that at or below v_1. Several polymers show even greater effects than does nylon.

In lubricated systems the starting friction is often higher than the kinetic friction. When the surfaces slide, lubricant is dragged into the contact region and separates the surfaces. This will initially lower the coefficient of friction, but at a still higher sliding speed the viscous drag increases as does the coefficient of friction as shown in Figure 6.31 and discussed more completely in Chapter 7 on *Lubrication*. This McKee-Petroff curve is typical for a shaft rotating in a sleeve bearing. The abscissa is given in units of ZN/P where Z is the viscosity of the lubricant, N is the shaft rotating speed, and P is the load transferred radially from the shaft to the bearing. (In the case of reader heads on magnetic recording media, the starting friction is referred to as "stiction.")

One source of apparent stick-slip (discussed further in *Analysis of Strip Chart Data,* later in this chapter) may arise from molecularly thin films of liquid. Static and flat bodies, between which is a thin layer of lubricant, induce crystalline order in the liquid. Then with motion of one plate there are periodic shear-melting transitions and recrystallization of the film. Uniform motion occurs at high velocity where the film no longer has time to order itself.

A frequent consequence of a static friction that exceeds kinetic friction is system vibration, which is discussed in a following section titled *Testing*.

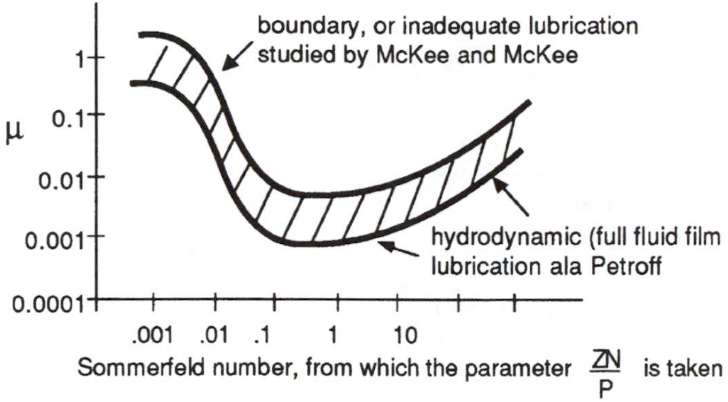

Figure 6.31 McKee-Petroff combined curves.

TABLES OF COEFFICIENT OF FRICTION

The coefficient of friction is not an intrinsic property of a material or combinations of materials. Rather it varies with changes in humidity, gas pressure, temperature, sliding speed, and contact pressure. It is different for each lubricant, for each surface quality, and for each shape of contact region. Furthermore, it changes with time of rubbing, and with different duty cycles. Very few materials and combinations have been tested over more than three or four variables, and then they are usually tested in laboratories using simple geometries. Thus, it is rarely realistic to use a general table of values of coefficient of friction as a source of design data. Information in the tables may provide guidelines, but where a significant investment will be made or high reliability must be achieved, the friction should be measured using a prototype device under design conditions.

Figure 6.32 is a graphical representation of coefficient of friction for various materials showing realistic (and usually disconcerting) *ranges* of values. A major deficiency in Figure 6.32 and all tabular forms is that they cannot show that friction is rarely smooth or steady over long periods, repeatable, or single valued.

VIBRATIONS AND FRICTION

No mechanical sliding system functions perfectly smoothly. They often vibrate, as may be seen when measuring friction forces. Most vibrations are benign, perhaps producing some audible sound. Sometimes, however, the vibrations are of such amplitude and frequency as to annoy people. Examples are brakes, clutches, sport shoes on polished floors, bearings in small electric motors, cutting tools, and many more. (Musical instruments that require the bow also emit sound but usually of a desirable nature.) The more extreme vibrations may

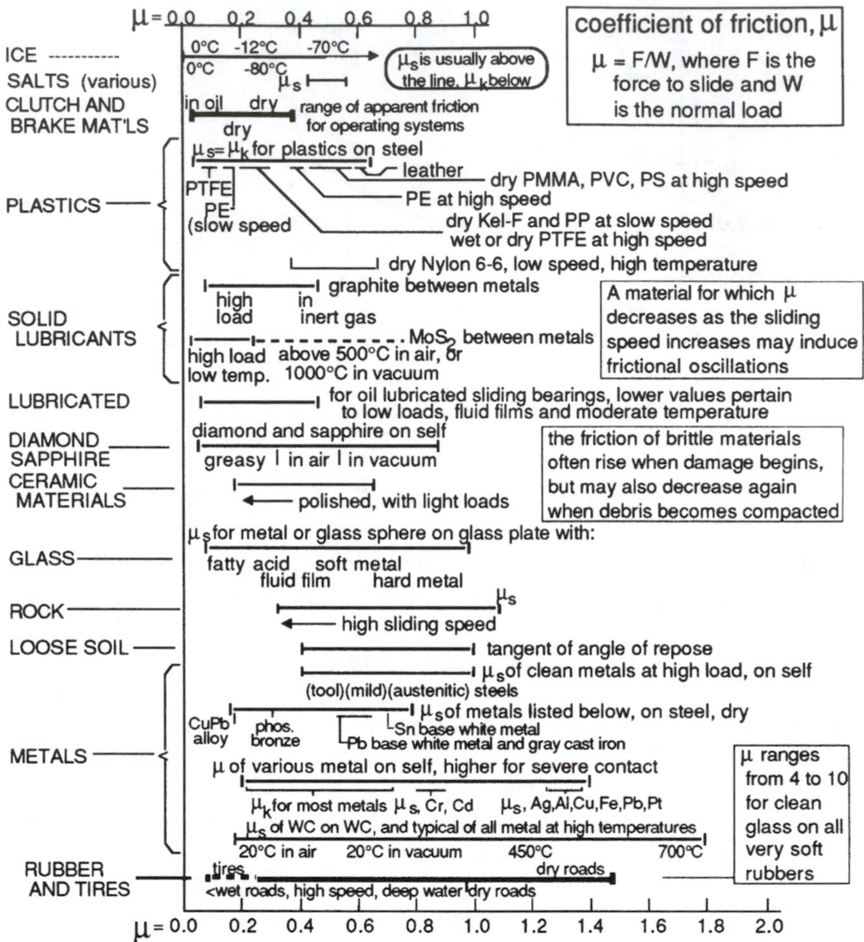

Figure 6.32 Some values of the coefficient of friction for various materials.

even damage machinery or in manufacturing processes may produce useless parts. Perhaps the most distressing part of frictional vibrations from the point of view of product designers is that there is no simple analytical method whereby frictional vibrations may be predicted.

Frictional vibration is an important problem in the measurement of friction and wear. Many investigators, have found that the consequence of vibration is a change in the (measured) friction, usually a reduction, but not always. Under some conditions the wear rate is affected as well, sometimes increasing it and sometimes decreasing it.

Frictional vibrations in machinery result from *both* the dynamics of the mechanical system holding the sliding pair *and* from the frictional properties of the materials that are sliding. This statement must be so because frictional vibrations can usually be *stopped* by changing slider materials or *reduced* by altering the mechanical system. This is a topic in which very strong biases appear among the specialists in dynamics and materials.

Research on frictional *behavior* of materials is usually empirical in nature since there is not yet a fundamental understanding of relevant frictional properties of materials. Part of the problem is that friction is not usually measured in a manner to determine potential vibration-inducing mechanisms. Most testing is done by rather arbitrary designs of test geometry, and the researcher hopes to achieve steady-state sliding, apparently on the assumption that steady-state sliding is the base condition of sliding.

Research on mechanical dynamics by contrast is quite mathematical because the (very) few fundamentals are well understood. Some research in this area is done by working backward from machine behavior to infer the frictional behavior of the sliding surfaces. The materials for the experimental phase of that research are usually not well chosen from the point of view of known frictional behavior. After the data are analyzed, a frictional model for the materials is often proposed as if the basic characteristic of the material had been found. Surely, the derived frictional model is strongly dependent on the mechanical model chosen for the mechanical system. There is no way to verify these results because there is no independent method of characterizing the frictional behavior of the materials in vibration conditions. We therefore see a dichotomy in published papers on frictional vibrations. *Published information on the frictional behavior of materials presumes the steady state and is not directly applicable to research on frictional vibrations, whereas the results of research on frictional vibrations appear to show very different frictional properties which are not possible to verify by conventional friction tests.*

One expectation in research on frictional vibrations is that a sliding speed or some other condition may be found at which frictional vibrations cease or do not exist. Such conditions may be calculated in nonlinear and properly damped systems in which the driving force is known or readily characterized. However, in most sliding systems the driving force (variations in friction) is usually not well known, or must be derived from a simulative test. It is possible that frictional behavior of a material may change over a range of sliding speed to eliminate frictional vibrations, but this cannot be predicted from machine dynamics alone. At best then, frictional vibrations might be reduced to an acceptable amplitude by changes in system dynamics, or its frequency may be moved out of unacceptable ranges.

The tendency for a sliding system to initiate/sustain frictional vibrations depends on the sensitivity of the mechanical system to vibrate in response to the frictional behavior of the sliding materials (including lubricants). These topics will be discussed in the section titled *Testing*.

Effect of Severe Uncoupled Vibration on Apparent Friction

Bolts in vibrating machinery and objects on vibrating tables often appear to move much more readily than if ordinary friction forces were operative. One explanation is that the two contacting surfaces may be accelerating at different rates from each other in the plane of their mutual contact. Another explanation may be that the two bodies separate from each other for a small amount of time. This latter idea is supported by experiments using a vibrator, in particular an ultrasonic horn, oscillating at 20 kHz. It was mounted on a sine table with precision of 0.0001 inch over 10 inches, which corresponds to an angular accuracy of .0006°. The sine table was set at a particular angle and the horn was set into oscillation. The power to the ultrasonic transducer was increased until the specimen began to slide downhill. Each power setting of the transducer produced a different amplitude of vibration of the lower specimen surface. The data are sketched in Figure 6.33.

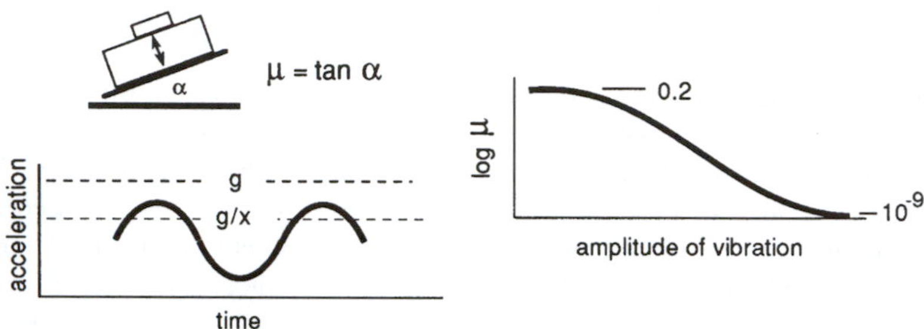

Figure 6.33 The effect of vibration on friction.

When the acceleration, a, of the vibrating surface exceeds the acceleration of gravity, g, there is complete momentary separation. When a = 0.9 g, there is very light contact for at least half of the cycle. Any attempted motion during contact probably involves elastic compliance which is released on the next half cycle.

Tapping and Jiggling to Reduce Friction Effects

One of the practices in the use of instruments is to tap and/or jiggle to obtain accurate readings. Tapping the face of a meter or gage probably causes the sliding surfaces in the gage to separate momentarily, reducing friction resistance to zero. The sliding surfaces (shafts in bearings or racks on gears) will advance some distance before contact between the surfaces is reestablished. Continued tapping will allow the surfaces to progress until the force to move the gage parts is reduced to zero.

Jiggling is best described by using the example of a shaft advanced axially through an O-ring. Such motion requires the application of a force to overcome friction. Rotation of the shaft also requires overcoming friction, but rotation reduces the force required to effect axial motion. In lubricated systems the mechanism may involve the formation of a thick fluid film between the shaft and the O-ring. In a dry system an explanation may be given in terms of *components* of *forces*. Frictional resistance force usually acts in the exact opposite direction of the direction of relative motion between sliding surfaces. If the shaft is rotated at a *moderate rate*, there will be very little frictional resistance to *resist* axial motion. In some devices the shaft is rotated in an oscillatory manner to avoid difficulties due to anisotropic (grooved) frictional behavior. Such oscillatory rotation is called jiggling, fiddling, or coaxing.

Jiggling, fiddling, and coaxing would appear to be anachronistic in this age of computer-based data acquisition systems. To some extent instruments are better and more precise than they were only 20 years ago, but it is instructive to tap transducer heads and other sensors now and then, even today.

TESTING

The effort of measuring friction can be avoided if one can find published data from the (near) exact material pair and sliding conditions under study. The exercise of measuring friction can be confusing because the data are almost never constant, rarely reproducible, and often confused by the dynamics of the measuring system. A first viewing of the usual irregular test results readily leads to doubt that the measurements were well done — but, that should rather cast doubts upon the neatness and simplicity of published values of friction, particularly those in tabular form!!

The difficulty in obtaining useful friction data may be seen in the exercise of formulating standards for friction test methods as by a committee of the American Society of Testing and Materials (ASTM). Several experienced people obtain identical test devices, identical materials and lubricants, identical data recording systems in some instances, and proceed to obtain data. The resulting data often differ by 25% or more leading to lengthy discussions on how to conduct further tests. Specimen preparation and other methods are revised and further testing is done. Often three or four iterations are required to obtain reasonable agreement of all data.

Standard test methods and accompanying test devices are useful for some commercial purposes, particularly when materials and mechanical components must meet certain specifications. However, having achieved a standard testing method it is often disconcerting to discover that the test conditions for achieving reproducibility are usually not those that accord with practical situations: they rarely simulate real or practical systems sufficiently.

The irregularity of data from laboratory test devices is also seen in the behavior of most practical sliding members. There are generally three reasons:

a. Sliding materials are inhomogeneous and their surfaces are rough at the start of sliding, and even more so after some sliding and wearing.
b. All sliding systems, practical machinery and laboratory devices, vibrate and move in an unsteady manner because of their mechanical dynamics.
c. Instrumented sliding systems will show behavior in the data that is affected by the dynamics of amplifier/recorders.

Measuring Systems

Measurement of the coefficient of friction involves two quantities, namely F, the force required to initiate and/or sustain sliding, and N, the normal force holding two surfaces together.

a. Simple devices: Some of the earliest measurements of the coefficient of friction were done by an arrangement of pulleys and weights as shown in Figure 6.34. Weight P is applied until sliding begins and one obtains the static, or starting, coefficient of friction with $\mu_s = P_s/N$. If the kinetic coefficient of friction μ_k is desired, a weight is applied to the string, and the slider is moved manually and released. If sliding ceases, more weight is applied to the string for a new trial until sustained sliding of uniform velocity is observed. In this case, the final weight P_k is used to obtain $\mu_k = P_k/N$.

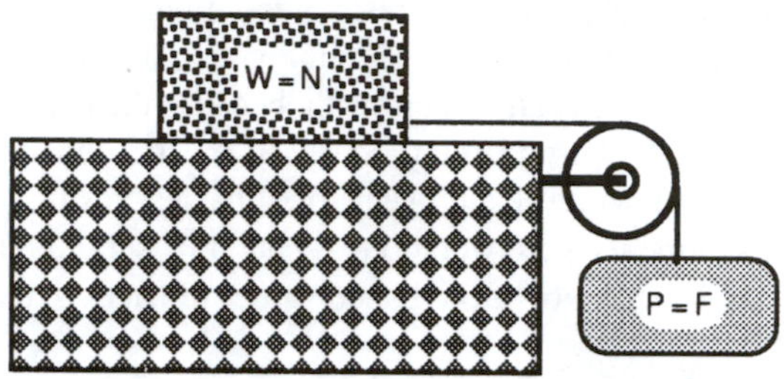

Figure 6.34 Dead load method of measuring friction.

A second convenient system for measuring friction is the inclined plane shown in Figure 6.35. The measurement of the static coefficient of friction simply consists of increasing the angle of tilt of the plane to α when the object begins to slide down the inclined plane. If the kinetic coefficient of friction is required, the plane is tilted and the slider is advanced manually. When an angle, α, is found at which sustained sliding of uniform velocity occurs, tan α is the operative kinetic coefficient of friction.

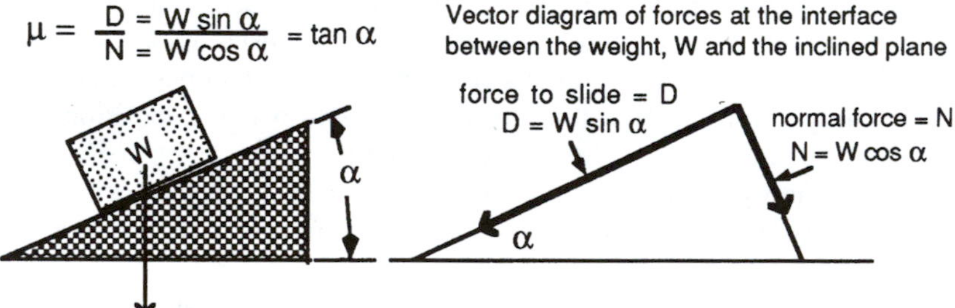

Figure 6.35 Slippery slope method of measuring friction.

b. Force measuring devices: As technology developed, it became possible to measure the coefficient of friction to high accuracy under dynamic conditions. Force measuring devices for this purpose range from the simple spring scale to devices that produce an electrical signal in proportion to an applied force. The deflection of a part with forces applied can be measured by strain gauges, capacitance sensors, inductance sensors, piezoelectric materials, optical interference, moire fringes, light beam deflection, and several other methods. The most widely used, because of simplicity, reliability, and ease of calibration, is the strain gage system. Others are more sensitive and can be applied to much stiffer transducers.

Just as there are many sensing systems available, there are also many designs of friction measuring machines. All friction measuring machines can be classified in terms of their vibration characteristics as well as range of load, sliding speed, etc. Only the pin-on-disk geometry will be discussed here, where the pin is held by a cantilever-shaped force transducer. While the pin-on-disk geometry is rarely a good simulator of practical devices, it is the most widely used configuration in both academic and industrial laboratories.

The principles of the interaction between cantilever vibrational properties and the frictional properties of the sliding pair may be illustrated by use of Figure 6.36, for a fixed root (not hinged) transducer. The prime mover moves as shown and the specimen plate offers resistance, F, to the sliding movement of the upper specimen, the pin. The cantilever bends backward, which can be measured by strain gages applied near the root of the cantilever on the vertical surface. The vertical force upon the upper slider can be measured by strain gages applied near the root of the cantilever on the horizontal (upper and lower) surface.

(See Problem Set questions 6 f and g.)

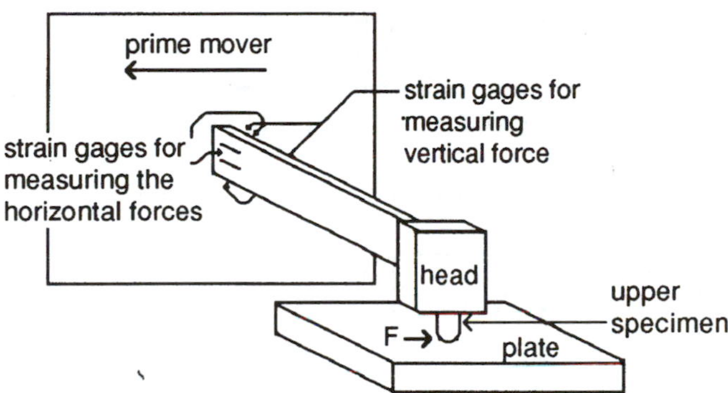

Figure 6.36 Sketch of a cantilever transducer for measuring friction force.

Force F is not coincident with the horizontal centerline of the cantilever, as shown in Figure 6.37, which is a view of the head of the transducer. A friction force thus applies a moment to the cantilever. When the upper slider is in the leading position relative to the vertical axis of the transducer, a frictional impulse rotates the transducer, simultaneously imparting a lifting impulse to the transducer and increasing the vertical load on the sliding contact region. This action constitutes a coupling between the vertical and horizontal mode of deflection of the transducer. By contrast, a frictional impulse upon a slider in the lagging position will also couple the vertical and horizontal deflection modes but in the opposite direction. When the slider is in the middle position a small impulse would produce very little coupling.

A better position for the sliding end of the upper slider would be coincident with both the vertical and horizontal centerline of the transducer. It is also possible to place the point of sliding contact above the horizontal centerline, in which case the leading position would act like the lagging position for the sliding contact

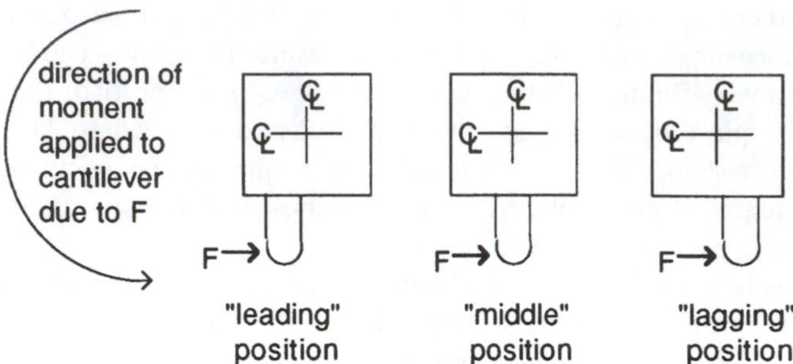

Figure 6.37 Sketch of the three common positions of the upper slider relative to the vertical and horizontal centerlines of the cantilever transducer.

point below center. Again there would be coupling between the vertical and horizontal deflection modes of the transducer.

A second type of coupling may occur when a transducer is tilted in the manner shown in Figure 6.38, and the stiffness in the horizontal direction is lower than that in the vertical direction. When the prime mover moves in the direction shown, a friction force, F, will be exerted that will bend the transducer in a direction having an upward component, by an amount dependent on the angle ε. In the case shown, a friction force will have the effect of reducing the vertical load on the sliding contact. When ε is in the opposite sense, a friction force will have the effect of increasing the vertical load on the sliding contact.

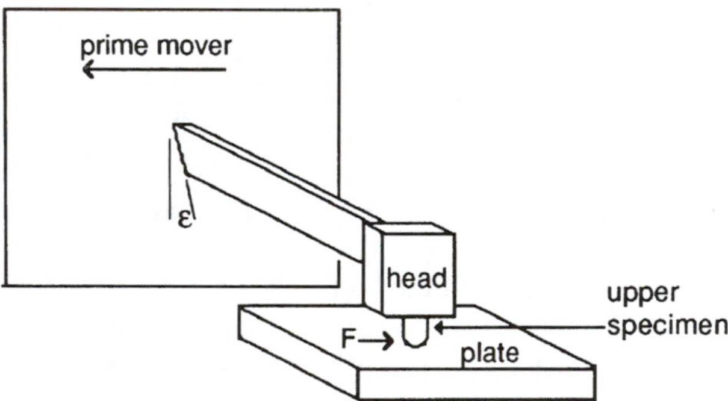

Figure 6.38 Sketch of a cantilever transducer that was oriented, in construction, at an angle ε relative to the vertical.

Static coupling of forces is virtually eliminated in the hinged cantilever transducer system sketched in Figure 6.39. The load is not applied by bending the cantilever in the vertical direction, but rather a load is applied in some manner directly upon the cantilever or head. Either a mass or a force can be applied anywhere along the cantilever, or upon an extension of the cantilever beyond the head.

At low sliding speeds where the upper slider may follow the contours of the plate there is no significant change in applied contact pressure. However, there may

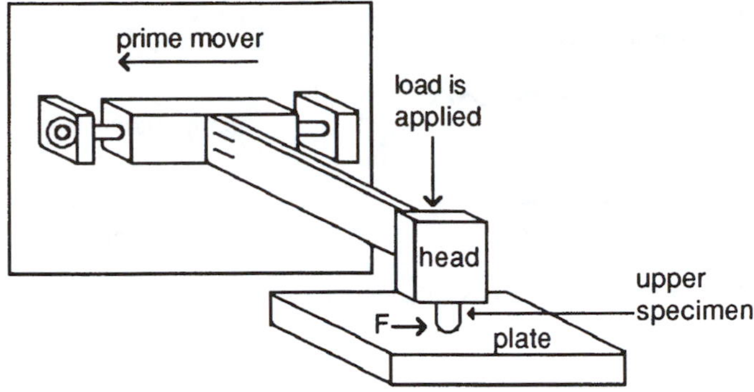

Figure 6.39 The hinged cantilever transducer system.

be some coupling between the vertical and horizontal forces when the slider, having some mass, moves at a higher speed as will be shown a little later in Figure 6.43. This effect will be maximum when the dead load is firmly attached above the pin, reduced when the mass is connected to the upper specimen through a weak spring, and reduced still more when loading is applied with an air cylinder. Note that again there may be some coupling where the axis of the hinge is not parallel with the plate. There will be no coupling where the flat bar cantilever in Figure 6.39 is tilted slightly although there may be small measurement errors.

(See Problem Set questions 6 h, i, and j.)

Interaction Between Frictional Behavior and Transducer Response

The three cantilever transducers in Figures 6.36, 6.38, and 6.39 are shown to be very flexible (compliant). If the bar is 1/4 inch thick, 1 inch wide, and 10 inches long (and held rigidly at its root) and the head is a 1-inch cube, both in steel, the horizontal natural frequency is about 50.5 Hz. The vertical natural frequency of the bars in Figures 6.36 and 6.38 will be about 202 Hz. (A 2-inch square bar 5 inches long would have a natural frequency in both directions of about 26 kHz. The force at the end of such a stiff bar would probably not be resolvable with strain gages, and may require the measurement of the deflection of the end of the bar by an inductive sensor or optical interference sensor.)

Several types of inherent frictional behavior can initiate and sustain vibration of the transducer during sliding. For example, the friction (as measured by some ideal system) might vary as shown in Figure 6.40. Upon sliding a pin over such a material the varying friction force constitutes a forcing function upon the cantilever. The variation in μ sketched in Figure 6.40 contains several frequencies which can be separated by Fourier analysis. Some of these frequencies will be below and some above the several natural frequencies of the transducers (and other parts of sliding machinery).

As a transducer vibrates in the horizontal direction the sliding velocity varies. If the friction (see Figure 6.41) decreases as the sliding speed increases there is a positive feedback with an increase of vibration amplitude, and vice versa.

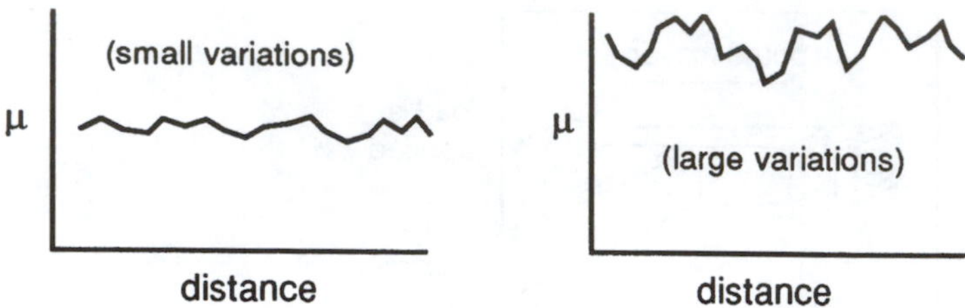

Figure 6.40 The variations in the coefficient of friction, μ, during sliding for two different materials.

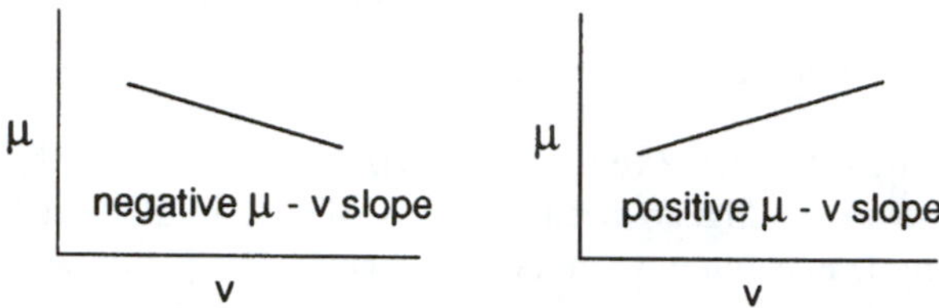

Figure 6.41 Two simplified variations of μ versus sliding speed, v.

There is virtually always some cross coupling between the six degrees of freedom of a transducer. That is, a vertical oscillation (and other modes) of the pin will usually accompany any varying horizontal friction forces during sliding. This action may be referred to as vertical-horizontal coupling and occurs even where friction is independent of contact pressure and sliding speed. The resulting variation in vertical force may produce variations in friction as shown in Figure 6.42, resulting in either an increase or decrease in vibration amplitude of the system.

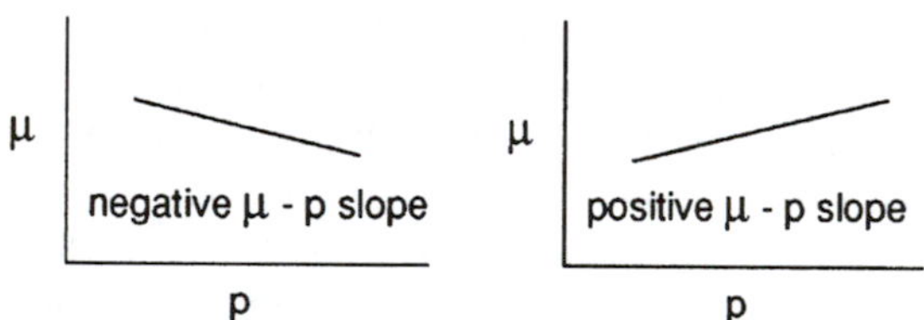

Figure 6.42 Two simplified variations of μ versus contact pressure, p.

Vertical-horizontal coupling could arise from:

a. Plastic flattening of asperities that plastically deform upon compression/traction contact
b. Rising of asperities that elastically strain upon traction/compression contact
c. Surface roughness that is greater than the effects of the two above stated effects
d. "Hot spots" — local regions that heat and expand and "lift" the counter-surface away.

The nature of surface coupling may change as speeds increase, due to jumping or hammering as shown in Figure 6.43. These phenomena have the effect of providing impulses in both contact pressure and sliding speed, which may have their separate effects on μ. These effects should be greatest in the *fixed root* transducer at low speeds, and greatest in the *hinged* transducer at high speeds when loaded directly with a mass.

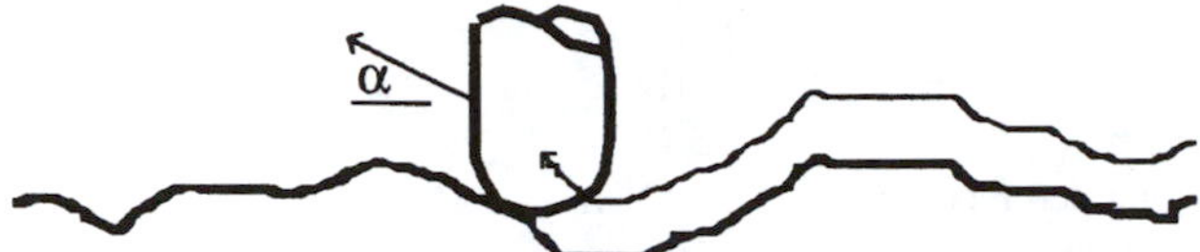

Path of the slider at low speeds: follows surface contour

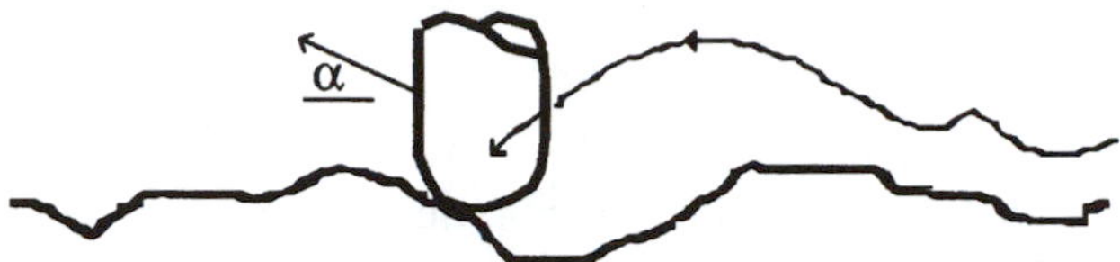

Path of the slider at high speeds, showing inertial lift-off

Figure 6.43 Sketch showing how vertical-horizontal coupling of motion may be affected by sliding speed.

In some instances friction changes gradually after a change of such variables as shown in Figure 6.44, which shows that friction may not change immediately upon changing sliding speed, load, or other variable. This effect can cause some confusion where the sliding speed varies over intervals of time less than the period of the friction transient.

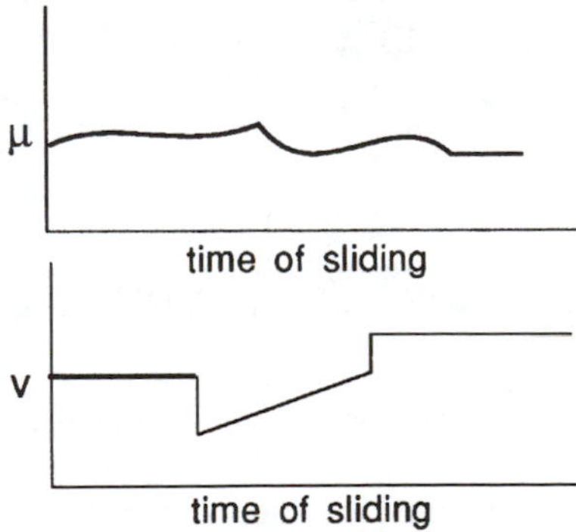

Figure 6.44 Delayed frictional changes when sliding speed changes.

Electrical and Mechanical Dynamics of Amplifier/Recorders

The electrical and mechanical dynamics of amplifier/recorders (data conditioning/acquisition systems) alter the information received from transducers. In some cases the "d.c." component is affected by the time-varying component, presenting false steady-state values. Amplifiers and recorders all have natural frequencies (and internal damping), approximately as follows (midrange values):

a. Voltmeters	1Hz
b. Pen recorders	5 Hz
c. Ultraviolet pen recorders	100 Hz
d. Computer-based	10 KHz

Data on the actual performance of each should be obtained from manufacturers. ***(See Problem Set question 6 k)***

Damping

Amplifiers and recorders alter the amplitude of input signals according to the match between the frequency of dynamic input signals and the natural frequencies of the amplifiers and recorders. Where they match, the output is large; where the input frequency is larger than the natural frequency, the signal will be altered in phase. Further, damping at various points in the system will affect the output. It is instructive to observe the simple series springs-masses-dashpots sketched in Figure 6.45. The system output may totally obscure the nature of the input.

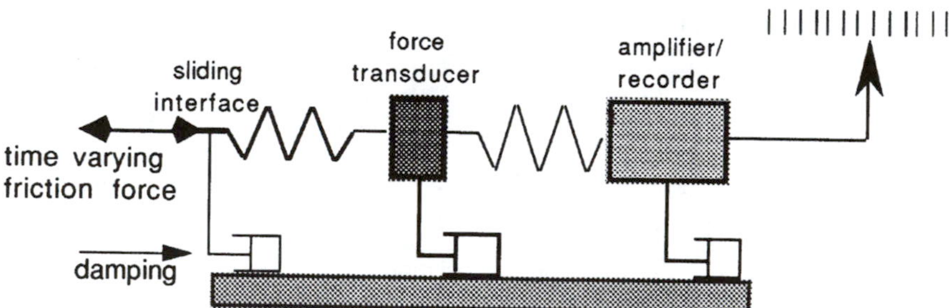

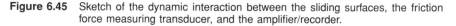

Figure 6.45 Sketch of the dynamic interaction between the sliding surfaces, the friction force measuring transducer, and the amplifier/recorder.

Friction often varies with time of sliding and even after time of standing between tests. Variations have been traced to wear and other changes of surfaces, and chemical changes.

ANALYSIS OF STRIP CHART DATA

Data obtained from friction measuring devices are usually not easy to interpret. For some sliding pairs a smooth force trace *may* be obtained on recorder strip chart but most often the friction force will drift or wander inexplicably. In other tests, where a flat plate rotates under a stationary pin, for example, variations in excess of 10 or 20% of the average force trace may be found during repeat rotations of the flat plate. These variations are often explained in terms of the stochastic nature of friction, but close examination will show real causes, such as spatial or temporal variations in surface chemistry, and wear. Variations are usually largest with small normal loads and are reduced at high loads, where contact pressures approach the state of fully developed plastic flow.

Vibration during sliding is often quickly referred to as "stick-slip." Laboratory devices can indeed be made to demonstrate true stick-slip, that is, alternating fast motion and stopping. The data from such an experiment will have the appearance of Figures 6.46a and 6.46b. Such behavior is rare in engineering practice. Usually, vibratory sliding can be better described in terms of Figures 6.46c and 6.46d. These figures show the velocity of a slider and the force applied to the slider by the prime mover.

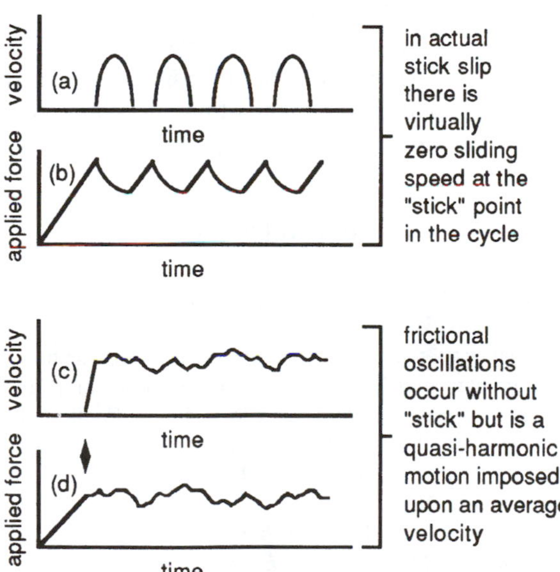
velocity
applied force
(a)
(b)
time
time
in actual stick slip there is virtually zero sliding speed at the "stick" point in the cycle
velocity
applied force
(c)
(d)
time
time
frictional oscillations occur without "stick" but is a quasi-harmonic motion imposed upon an average velocity

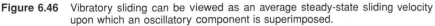

Figure 6.46 Vibratory sliding can be viewed as an average steady-state sliding velocity upon which an oscillatory component is superimposed.

The value of μ_s may be obtained from the maximum force measured when slip starts, as indicated by the arrow in Figure 6.46d. The shape of the curve prior to the maximum reflects only the system stiffness and speed of the prime mover.

When slip begins, the "slip" portion is usually not recorded in sufficient detail to determine μ_k. In general, it is incorrect to assume that μ_k is the average of peaks and minima in the excursions because in traces such as those shown in Figure 6.46a, μ_k would be approximately equal to $\mu_s/2$.

In the more common trace for small oscillations as shown in Figure 6.46c, μ_k may be taken as the average of the trace. Where excursions are greater than about 20% of the midpoint, value averaging must be done with caution. It is better to damp the oscillation of the machine than to average the traces from a severely vibrating machine, even though damping will likely alter the dynamics of the system.

HOW TO USE TEST DATA

It is best to measure friction of contacting pairs in practical conditions, including the vibrations, time of standing still between uses, varying sliding speed, etc. If measurements are to be done in a laboratory, they should be done on a test device and in the manner that closely simulates the full range of variability of the practical environment, including various states of wear or surface change due to sustained use. There is little point in attempting to measure friction (or wear rate) in steady-state sliding because there is no reliable way to connect the data to any unsteady-state sliding conditions.

When data are obtained it is not useful to record average values or steady-state values of friction coefficient, but rather the range of values should be noted together with some description of the nature of unsteadiness and the time varying trends. Test data reflect reality; research papers and books less so.

REFERENCES

1. Bowden, F.P. and Tabor, D., *Friction and Lubrication of Solids*, Oxford University Press, Oxford, U.K., Vol. 1, 1954; Vol. 2, 1964.
2. D. Dowson has written a most interesting series of biographies of 23 prominent people in tribology (mostly in lubrication). The series may be found in a book entitled, *The History of Tribology*, Longman, Essex, England, 1978. Sketches of the biographies may also be found in a series of issues of the *ASME Journal of Lubrication Technology* from Vol. 99 (1977) to Vol. 102 (1980).
3. Ludema, K.C, *Wear*, 53, 1, 1979.
4. Gane, N., *Proc. Roy. Soc.,* (London), A 317, 367, 1970.
5. Shaw, M.C. and Ber, M., *Trans. ASME*, 82, 342, 1960.
6. Grosch, K.A., *Proc. Roy. Soc.*, (London), A 274, 21, 1963.
7. Schallamach, A., *Wear*, 6, 375, 1963.
8. Ludema, K.C and Tabor, D., *Wear*, 9, 439, 1966.
9. Bahadur, S., *Wear*, 18, 109, 1971.
10. Johnson, K.L., Kendall K., *Proc. Roy. Soc.,* (London), A. 324, 127, 1971.
11. Roberts, A.D., *J. Phys. D: Appl. Phys.*, 10, 1801, 1977; Mortimer, T.P. and Ludema, K.C., *Wear*, 28, 197, 1974.
12. Montgomery, R., *Wear*, 33, 359, 1975.
13. Barnes, P. and Tabor, D., *Proc. Roy. Soc.,* (London), A324, 127, 1971.

CHAPTER 7

Lubrication by Inert Fluids, Greases, and Solids

FLUID FILM LUBRICATION IS INDISPENSABLE FOR LONG LIFE OF HIGH SPEED BEARINGS, VERY USEFUL IN COMMON MACHINERY BUT OF LESSER INTEREST IN SIMPLER CONSUMER PRODUCTS. GREASES ARE ADEQUATE IN LOW SPEED MECHANISMS, WHERE LIQUID CIRCULATION IS NOT WARRANTED ECONOMICALLY. SOLID LUBRICANTS ARE USED IN HIGH TEMPERATURE AND EXTREME CONTACT PRESSURE APPLICATIONS, BUT USUALLY NOT FOR LONG PRODUCT LIFE. CHEMICALLY ACTIVE CONSTITUENTS IN LUBRICANTS ARE DISCUSSED IN CHAPTER 9.

INTRODUCTION

Sliding surfaces in the home are often lubricated to stop them from squeaking, or to make them last longer. Machine bearings are lubricated in order to prevent seizure and to achieve a long life. In the 20th century, friction reduction has been of lesser concern than seizure or wear, but friction was important in the 18th century when animal power was most widely applied and in the 19th century when railroads were being developed. It has become important again as the cost of fuel has risen, a trend that began in the early 1970s.

Bearings are designed to meet certain requirements, usually expressed in terms of load carrying capacity, stiffness, and dynamic behavior. Many of these properties are quantified, but good design also involves several nonmathematical variables, such as how the lubricant is applied, how to accommodate misalignment, and what to do about starting and stopping a bearing.

FUNDAMENTAL CONTACT CONDITION AND SOLUTION

The primary objective in lubrication is to reduce the severity of both the normal and shear stresses in solid surface contact. One universal fact in the theories of friction and wear is that only a small fraction of the nominal area of

The work of all named authors in this chapter is described in references 1,2 unless specifically cited.

contact between two bodies is in actual contact. The actual contact area may be as little as 0.01% of the apparent area of contact, and no stresses exist in the regions between. The stress-state in most of the actual areas of contact exceeds the yield point of ductile materials and the fracture strength of brittle materials. All of the mechanical energy applied to (absorbed by) an unlubricated bearing heats and deforms the sliding surfaces.

The adverse effects of contact between rough surfaces can be reduced by smoothing out the variations in surface stress to some lower average values. This smoothing may be accomplished by inserting a film of compliant material between solid surfaces, such as a pad of rubber. However, a pad of rubber cannot be readily accommodated between moving parts in a machine.

PRACTICAL SOLUTION

Liquids and soft solids are effective lubricants: the range is unlimited and includes gasoline, mercury, catsup, acids, mashed potatoes, and oil in refrigerant. Suet and other organic matter served as sufficient lubricants until the last century, at least for slow machines. Suet ultimately was inadequate to the task, yielding to pumpable fluids and more socially acceptable grease. In the 1930s, the simple fluid lubricants became the limit to some technological progress, and chemical additives were developed to improve lubrication. At about the same time, graphite and MoS_2 became well known both as additives to oil and for use without oil.

Proper design in the old days consisted of making bearings such that all available lubricant found its way to the critical regions, preferably by gravity such as in Conestoga wagon wheel bearings. (The wheels of Conestoga wagons rotated on stationary shafts. Thus the region of contact between the wheel hub and axle was at the bottom of the axle, which is where lubricant settled. Railroad car axles, by contrast, rotate in stationary bearings [journals] where the contact region is at the top of the axle.)

With the development of labor-saving machinery, more output was also expected from machines, and they were designed to carry larger loads and move even faster. The subject of lubrication is not readily outlined without ambiguity. However, the most common categories of lubrication are liquid film lubrication, boundary lubrication, and solid lubrication. These categories will be discussed in turn.

CLASSIFICATION OF INERT LIQUID LUBRICANT FILMS

Fluid films can be provided in a bearing, by:

1. Retention of a fluid in a gap by surface tension
2. Pumping fluid into a contact region (called hydrostatic lubrication)
3. Hydrodynamic action.

Surface Tension

If a drop of liquid is placed upon a flat surface and then another flat surface is laid upon the wetted surface, some liquid will be squeezed out, but not all of it. Surface tension, the same force that makes liquid rise in a very small diameter glass tube, will make complete exclusion of liquid very difficult. The amount that will be retained in the gap between two surfaces is related to the wettability of the liquid (lubricant) on the surface of interest. Wettability may be defined in terms of the contact angle, β, as shown in Figure 7.1.

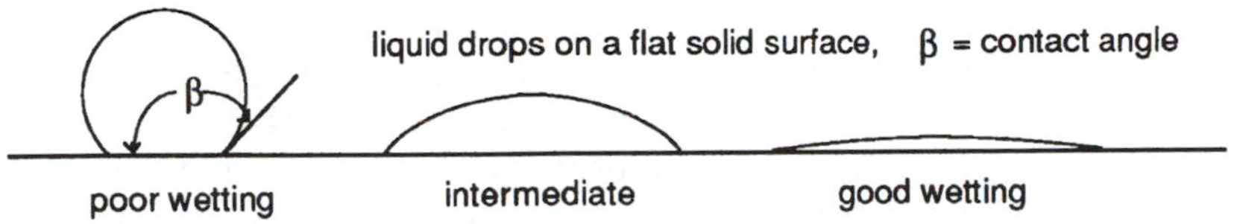

Figure 7.1 Contact angle related to wettability.

The contact angle of four common liquids on glass is given in Table 7.1.

Table 7.1 Contact Angle of Various Liquids on Glass

	β
H_2O	110°
H_2O + soap	80°
Furfural	30°
Isopropynol	<1°

If a drop of the lubricant spreads out completely and spontaneously on the surface, then most of that lubricant will also run out of the bearing. If the drop of lubricant stands up on the bearing surface as water does on a waxed surface, that lubricant will not readily enter the narrow contact regions of the bearing. If the drop has a base diameter about twice the height of the drop, the lubricant will enter the vital region and much of it will remain there. In the absence of a useful theory for molecular film lubrication, the drop spreading test is convenient for selecting materials and lubricants for applications where small quantities of lubricant are applied "for the life of the product."

A related phenomenon is capillary action, which is the basis for wick lubrication. The wick is a porous material (e.g., cloth) which has its lower end dipped in oil and its upper end in contact with the rotating shaft.

(See Problem Set question 7 a.)

Hydrostatics

Two sliding surfaces can be separated by pumping a fluid into the contact region at a sufficient pressure to separate the surfaces. A large volume of fluid will separate the sliding surfaces a great distance, thereby producing a low resistance to sliding motion. However, the energy required to pump the fluid must be considered in the overall economy of the bearing system. Hydrostatic lubrication is effective over all sliding speeds, but its reliability is influenced by the reliability of the required external pump.

Hydrodynamics

If one surface slides along another at moderately high speed, and if the shape of the leading edge of the moving surface is such that fluid can be gathered under the sliding surface, the two surfaces will be separated and slide easily. This is hydrodynamic lubrication. Water skiing operates on this principle, and it may be recalled that a major aspect of this sport is getting started. Hydrodynamics has been very thoroughly studied because of its practical significance. Very many books and technical papers are available on the subject, from the very mathematical to the very practical. Only a short summary is given on the following pages.

SHAFT LUBRICATION

The lubrication of shafts in sleeve or journal bearings has been widely studied in the last two centuries because these components are so widely used in power generating machinery and railroad equipment. (Strictly, a journal is "that portion of a rotating shaft, axle, spindle, etc., which rotates in a bearing." The stationary member is called a journal bearing.) G. Hirn was one of the early investigators of the behavior of these components. He lubricated some bearings with animal, vegetable, and mineral oils, and noted that the coefficient of friction, μ, was directly proportional to speed at constant temperature and was also directly proportional to viscosity of the lubricant. N. Petroff did the same, using Caucasian mineral oil in railroad axles.[2] He concluded that he was not measuring real friction, but a sliding resistance due to an intermediary layer. He called it "mediate friction," which was later interpreted to mean viscous drag.

The magnitude of viscous drag force for a fluid film between two parallel surfaces can be calculated with the equation given in Figure 7.2. This equation defines dynamic viscosity denoted by η. (There are many definitions and types of viscosity, which the reader may find in textbooks on fluid mechanics or lubrication.)

The units on dynamic viscosity may readily be recalled by using the sketch, where two blocks are separated by a fluid film of thickness, h and viscosity, η. The force, F, required to slide the upper block is: $F = \frac{\eta A v}{h}$

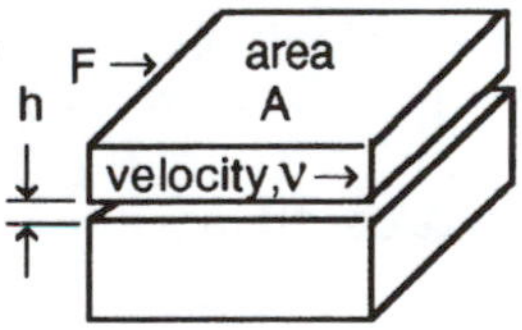

Figure 7.2 Definition of dynamic viscosity.

The units on η are M/LT. Viscosity is hereby defined functionally. Newton defined it as, "the resistance which arises from lack of slipperiness in a fluid..."

Strictly, the above definition applies to values of h greater than 50 times the dimension of the molecules in the fluid. Very thin films have higher viscosities, and films of the order of 5 nm thick begin to display solid properties.

Petroff calculated friction force, F, in lubricated bearings as the viscous drag of fluid in the (nonuniform) radial clearance space, c, between a shaft rotating in the center of a bearing, with a surface velocity of U. The wetted area is πDL where L is the length of the bearing, and D is the diameter of the shaft. Then,

$$F = \frac{\pi\eta DLU}{c} \tag{1}$$

from which μ can be calculated as F/W (where W is the applied load). This is Petroff's Law.

In 1883, B. Tower presented the results of a study of bearing friction. He used a 6-inch-long half bearing on a 4-inch-diameter shaft with 180° conformity. The shaft was immersed in oil and rotated, with a load of 8008 pounds applied to the bearing. He measured the hydraulic pressure at various locations in the thin space between the shaft and bearing. The pressure peaked sharply behind the center of contact. By integration over the 180°, Tower calculated that the film was carrying a load of 7988 pounds. He verified that lubricant efficacy for a shaft rotating in a bearing varied with lubricant viscosity, bearing dimension, and machine speed as others had reported. Most important, he found that the large variations in reported friction were due to the varied methods of lubrication.

In 1886, O. Reynolds developed some equations for the case of the flooded (adequate lubricant supply) bearing with no flow of lubricant out the end of the bearing. He described the action of lubrication using the idea that the rotating shaft "drags" fluid into the contact region between itself and the bearing, building up a fluid pressure that carries the applied load. He combined these variables into a mathematical formulation based on the Navier-Stokes equations for fluid flow. Very many later authors used the Reynolds equations as the point of departure for their analysis of bearing behavior for such difficult cases as narrow bearings (considerable side leakage), high loading, and variations on conditions prevailing in the entrance wedge.

In 1904, A. Sommerfeld began publishing variations of the Reynolds equations for a number of practical conditions, particularly for the behavior of a shaft in a well-lubricated bearing. This case will now be described, with a note on the start-up of shaft rotation.

A stationary shaft of diameter, D, with a vertical load, W, in a bearing of inner diameter, D + 2c, is shown in Figure 7.3a. (c is the radial clearance.) As the shaft begins to rotate, it climbs one side of the bearing as shown in Figure 7.3b. If the shaft and the bearing are immersed in oil, the sliding shaft will drag oil underneath itself, to begin forming the hydrodynamic wedge. It is not a visible wedge since the entire system is immersed. Rather, it is a pressurized region which lifts the shaft. When the wedge is fully developed the shaft takes the position shown in Figure 7.3c, with a minimum separation, h. Note that the fluid film pressure builds *behind* the location of the minimum separation between the shaft and the bearing, taking the shaft surface as the reference. The bearing analysis community of the time considered certain variables to be convenient for discussion, and these included the eccentricity, ε, of the center of the shaft from the center of the bearing (defined as ε = 1 – h/c, and L/D, where L is the length of the bearing. A convenient formulation of the variables was:

$$\frac{\eta N}{p}\left(\frac{D}{c}\right)^2\left(\frac{L}{D}\right)^2 = \frac{\left(1-\varepsilon^2\right)^2}{\varepsilon\pi\sqrt{\left(1+.62\varepsilon^2\right)}} \tag{2}$$

where $p = \frac{W}{DL}$ and N is the rpm of the shaft.

The term on the left is one form of Sommerfeld's number and is sometimes referred to as the bearing characteristic. Bearings with the same characteristic will operate with the same eccentricity. This value is significant since it was found that for efficiency, h/c (which equals 1 – ε) should be about 0.3. The consequence of this recommendation would be a particular set of values for the adjustable variables ηN/p for a given bearing.

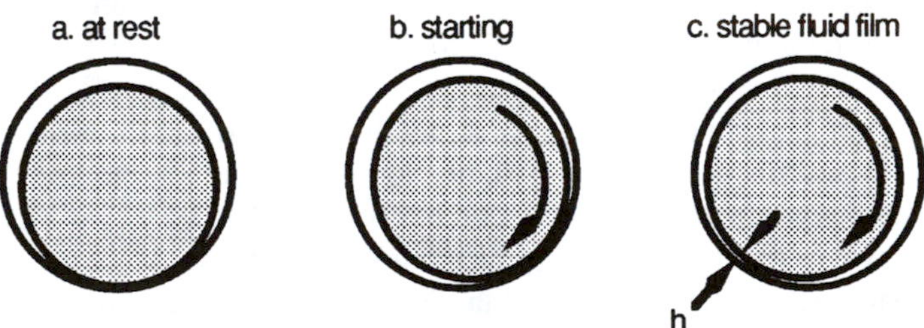

Figure 7.3 Three positions of a shaft in a bearing.

This same equation, with small variation, can be used to analyze bearings in which an unbalanced shaft rotates. If the static (vertical) load, W, on a horizontal shaft, is small as compared with an unbalanced force, the point of minimum lubricant film rotates with the shaft along the inner surface of the bearing. In this case the fluid wedge is ahead of the location of minimum film thickness. An interesting situation develops when an unbalanced shaft has a slightly larger and intermittent vertical load applied. The shaft will oscillate between rotating stably with the wedge behind the point of minimum separation, and circulating in the bearing with the wedge ahead of the point of minimum separation. In the transition between these two states, an existing wedge "collapses," leading to a thinner average fluid film and higher friction than for either stable condition. There are many cases of such instability but one in particular is where the shaft center will circulate within the bearing at half the shaft speed. This is referred to as shaft whirl, and a whirling shaft consumes more energy than does a stable shaft. Whirl is a problem in vertical shafts particularly.

R. Stribeck presented the results of a study of the friction of hydrodynamic bearings. He confirmed a minimum point in friction for a great number of varying conditions. L. Gumbel studied Stribeck's results and found that they could be unified into a single curve on coordinates μ versus $\eta\omega/p$. Hersey claimed to find the same convenient relationship, preferring shaft speed (rpm), N, to the angular velocity, ω, and Z in the place of η. ZN/p is the widely used quantity found on the abscissa on Stribeck curves (and ZN/p is sometimes referred to as the Hersey number). For completeness we should add considerations for side flow from the bearings and account for grooves in bearings.

There was a good analytical explanation of the bearing friction at higher values of ZN/p in Petroff's law, namely, it is due to viscous drag between well-separated solid surfaces. The McKee brothers located the minimum friction for a number of bearings by experiment.[3] It was widely agreed that at values of ZN/p less than that which produced minimum friction the lubricant film is thinner than the height of the asperities on the opposing metal surfaces. This condition is now referred to as "boundary lubrication," which is a misnomer (see the section titled *Scuffing and Boundary Lubrication* in Chapter 9). Typical data for a wide range of variables are shown in Figure 7.4.

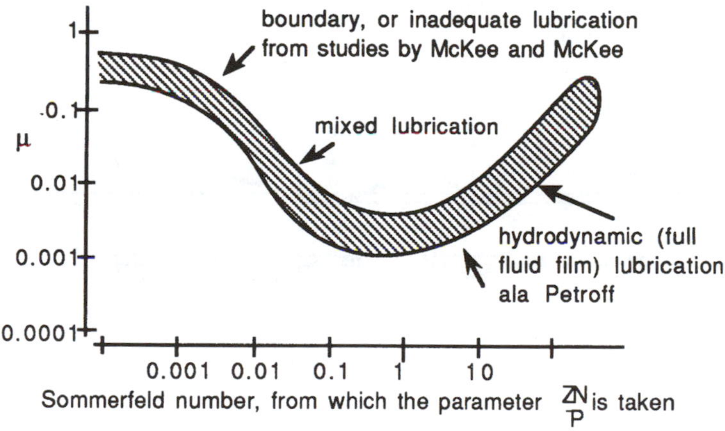

Figure 7.4 Stribeck-Gumbel or McKee-Petroff curves.

HYDRODYNAMICS

The Reynolds equations have been used by H.M. Martin as the basis for calculating the load carrying ability of gear teeth. The contact condition between gear teeth was simulated by edge contact of 2 discs of radius, R, and length, L, rolling against each other with an applied load of W and an average surface velocity of U:

$$\frac{W}{L} = 2.45\frac{UR\eta}{h_o} \tag{3}$$

The load-carrying capacity of a lubricant film is taken as the point at which h_o is so small that the tops of the asperities on the opposing surfaces begin to touch. Note that for disks rotating in the same direction, U = 0 and no film should develop: for two disks with equal surface velocity in the same direction the film should be the thickest of all conditions.

Though Martin did not first express this concept formally (because surface roughness was not adequately described until the surface roughness tracer was invented in 1936), the Λ ratio can be introduced here. Λ is defined as h_o/σ where σ is most often taken as $\sqrt{(\sigma_1^2 + \sigma_2^2)}$ and the σ values are the roughnesses of the two contacting surfaces expressed as the rms asperity height (thus σ should be expressed as R_z). Where $\Lambda > 1$ there is thought to be virtually no contact between asperities (even though σ is a statistical expression of asperity height) and thus little wear. (See Chapter 9 under *Friction in Marginal Lubrication.*) Figure 7.5 is a sketch which shows the locations of these quantities. Most researchers of that era were quite sure that calculated Λ was less than 1 for many successful machine components. Further it was noted by A.W. Burwell that "those oils least refined are, in general, better lubricants than the same oils highly refined."[4] There appeared to be a lubricating quality in oil therefore that was not explained in terms of viscosity. That quality was thought to be chemical in nature and will be taken up in Chapter 9.

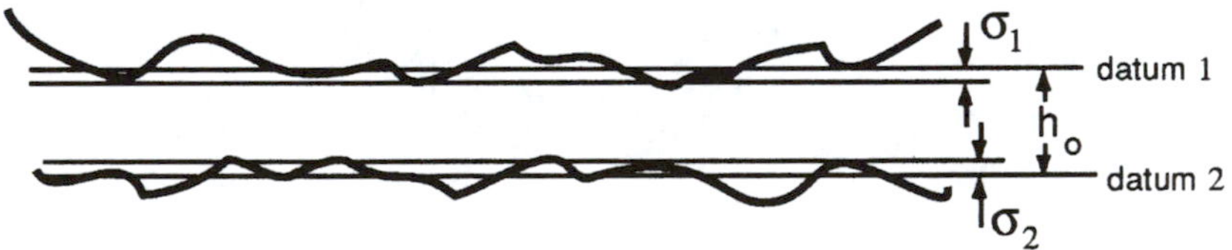

Figure 7.5 Sketch showing where surface roughness values and fluid film separation values are assumed to be.

However, close study showed that "oiliness" could not explain all of the limitations of Martin's equation, particularly at very high contact pressure between the discs and other components. Speculation on the exact nature of difficulty with the equation may be found in the literature of the 1930s and 1940s. The limitations of hydrodynamics were not a problem for most mechanical designers, many of whom recognized that the conservative equations rather nicely offset the poor dimensional tolerances to which many mechanical parts were made.

It was not until 1949 that A.M. Ertel of Russia showed the importance of elastic deformation in the region of contact. When a load is applied there is some elastic deformation of the surfaces, which increases conformity and broadens the region of close proximity of materials. The contact pressure is therefore lower, and an escaping fluid must traverse a greater distance than in the case of non-conforming contact, so the fluid film is thicker. Ertel had also incorporated a third effect into his analysis and that was the influence of pressure on increasing the viscosity of oil in the conjunction. Ertel's equation produced a film thickness (over most of the conjunction) that was about 10 times that of Martin and was widely accepted at once.*

Equations that combine both elastic and hydrodynamic considerations are known as elastohydrodynamic equations. There are many forms of ehd equations, depending on the adjustments one makes for mathematical convenience. They can only be solved accurately by numerical methods, and one such equation for edge contact of disks is due to D. Dowson and G. R. Higginson:[6]

$$\frac{h_{min}}{R'} = \frac{0.88(\alpha E')^{0.6}\left(\frac{\eta_o U}{E'R'}\right)^{0.7}}{\left(\frac{W}{LER'}\right)^{0.13}} \qquad \textbf{(4)}$$

where the effective plane strain Young's Modulus E′ is related to those of the two discs by

$$\frac{1}{E'} = \frac{1}{2}\left(\frac{1-\nu_1^2}{E_1} + \frac{1-\nu_2^2}{E_2}\right) \quad \text{where } \nu \text{ is the Poisson ratio}$$

and

$$\frac{1}{R'} = \frac{1}{R_1} + \frac{1}{R_2}$$

Several equations of nearly similar form are found in the literature, differing in coefficients and exponents mostly. These variations are a consequence of various geometries and assumptions in analysis and from the use of different databases in the empirically assisted equations. In these equations η_o is the bulk viscosity of the fluid as before, but account is taken of the increase in viscosity by pressure in the contact region by pressure viscosity index α (which has values for mineral oil in the region of 3×10^{-4} m^2/N). One difference to be noted from Martin's equation is that the minimum film thickness is denoted as h_{min} instead of h_o. The difference is due to a small projection of the contacting regions into the fluid film, as shown in Figure 7.6. Equations show a sharp peak in the fluid pressure

* Ertel was thought to have died in the great Soviet folly, but escaped to Germany, taking an assumed name. His work was salvaged from possible oblivion by his mentor, A.N. Grubin and was called the Grubin equation until Ertel felt secure enough to reveal himself.[5]

in the same region, which intuition would suggest should depress the materials in that region. However, the projection is about 25% of the average fluid film thickness and has been confirmed by experiment. It is important to verify the magnitude of a pressure spike: some of the higher published values are high enough to suggest that the most severe stress states in the substrate are much nearer the surface than 0.5a from Hertz equations. These stresses could induce fatigue failure in the surfaces of parts rather than in the substrate.

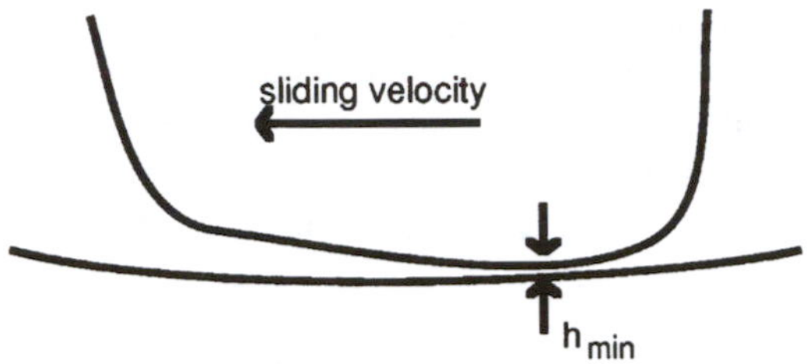

Figure 7.6 Sketch of elastohydrodynamic conjunction region.

A perspective on the conditions in the conjunction is given by Dr. L. D. Wedeven.[7] As a matter of scale, the conjunction has proportions such that the oil film is about ankle deep on a football field, and the viscosity of the oil is about like that of American cheese! Dr. Wedeven was the first to show the fluid film thickness distribution in the conjunction for a sphere sliding on a flat plate.

One enduring problem with fluid film lubrication is that bearings must be started from 0 velocity and occasionally have serious overloads applied or fall into a whirl. Another problem may be temporary starvation for oil, or a gradual decrease in the viscosity of the oil due to heating, such that the oil is no longer sufficient as a lubricant. In such cases certain chemical additives have been found to be useful. Since the additives appear to concentrate their influence at sliding boundaries, they are called boundary lubricants. (See Chapter 9.)

In bearing design there are at least three practical concerns. One is to impede the escape of pressurized lubricant from the conjunction: this requires fluid barriers at the end of the bearing, or long bearings, and requires proper location of lubricant feeder orifices and grooves. A second concern is the disposal of debris. If the debris has dimensions less than the fluid film it should produce little harm. A third concern is heat removal. Much heat is generated in the shearing fluid and some is generated in the solid surfaces when contact occurs. The lubricant is an agent for its removal. If heating occurs faster than does removal then a thermal spiral has begun, the lubricant degrades, and surfaces contact each other.

Current research in hydrodynamic lubrication focuses on the properties of fluids at high pressures, but particularly at high shear rates. There has been little success to date in predicting the friction or sliding resistance in thick-film lubrication.

(See Problem Set question 7 b.)

TIRE TRACTION ON WET ROADS

The friction of tires on dry roads was discussed in Chapter 6. Wet roads are actually lubricated surfaces to the tire. Figure 7.7 shows the effect of speed, wheel lock up, and amount of grooving (tread pattern) on the braking force coefficient. The braking force coefficient values given in Figure 7.7 were taken from two different tests with pavement of moderately polished roughness, with water equivalent to that which results from a moderate rainfall (as would require continuous windshield wiper motion at first speed). Polished road surfaces, thick water films from very heavy rain, and smooth tires reduce the braking force potential to values only a little higher than that of ice.

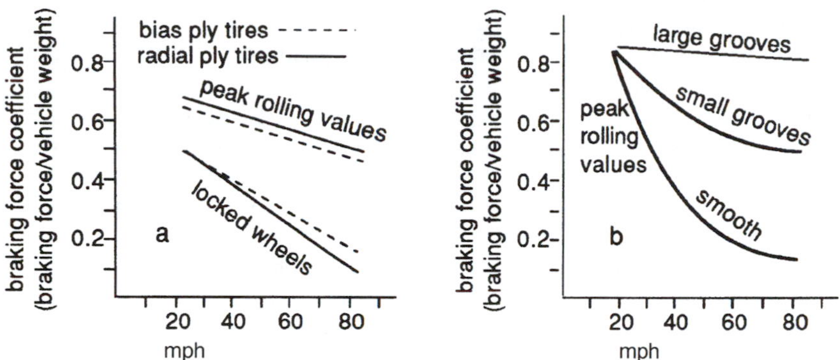

Figure 7.7 Results of two different tests of the skid resistance of tires on wet roads versus speed in miles per hour.

(See Problem Set questions 7 c and d.)

SQUEEZE FILM

When a shaft, tire, or skeletal joint (hips, etc.) stops sliding on a lubricant film, i.e., the velocity becomes zero, the equations of hydrodynamics would suggest the fluid film reduces to zero immediately. Actually there is a slight time delay, while the fluid squeezes out of the contact region. The time required can be estimated from the equation:

$$\frac{1}{h^2} = \frac{1}{h_o^2} + \frac{2W\left(a^2 + b^2\right)t}{3\pi a^3 b^3 \eta} \qquad (5)$$

for an elliptical-shaped contact of dimensions a and b, where h_o is the original film thickness (for small values of h_o relative to a or b), η is the dynamic viscosity of the fluid, and W is the load that produces a film of thickness h after time t.

For most steady-state engineering systems, the time to squeeze out a film is very small, on the order of milliseconds. For sliding surfaces, film thinning as speed decreases is much slower than the squeeze film effect. Fluid films do, however, cushion the impact striking surfaces in the presence of a fluid, as for a ball striking a surface or a shaft rattling within a sleeve bearing.

LUBRICATION WITH GREASE[8]

The word "grease" is derived from the early Latin word "crassus" meaning fat. Greases are primarily classified by their thickeners, the most common being metallic soaps. Others include polyurea and inorganic thickeners. Greases are usually not simply high viscosity liquids.

Soap-based greases are produced from three main ingredients:

1. The fluid (85–90% of the volume), which can be selected from mineral oils, various types of synthetics, polyglycols, or a never-ending combination of fluids.
2. A fatty material (animal or vegetable), which is usually 4 to 15% of the total, called the acid.
3. The base or alkali. Bases used in making greases include calcium, aluminum, sodium, barium, and lithium compounds, with 1 to 3% normally needed.

When a fat (acid) is cooked with the alkali (base), the process of forming soap by splitting the fat is known as saponification. When a fatty acid is used instead of a fat, the process is known as neutralization.

A more complex structure can be formed by using a complexing salt, thus converting the thickener to a soap–salt complex, hence the term "complex greases." Complex greases offer about a 38°C (100°F) higher working temperature than normal soap-thickened products. They were developed to improve the heat resistance of soap greases, the most popular being compounds of lithium, aluminum, calcium, and barium.

Inorganic thickeners, such as clays and silica (abrasive materials!!), consist of spheres and platelets that thicken fluids because of their large surface area. These products produce a very smooth nonmelting grease that can be made to perform very well when careful consideration is given to product application. Polyurea is a type of nonsoap thickener that is formed from urea derivatives, not a true polymer but a different chemical whose thickening structure is similar to soap. Polyurea greases are very stable, high-dropping-point (flow temperature) products that give outstanding service.

The lithium 12-hydroxystearate greases are by far the most popular. These are based on 12-hydroxystearate acid, a fatty acid that produces the best lithium and lithium complex grease.

Additives can impart certain characteristics that may be desirable in some cases. Extreme pressure (EP) and antiwear additives are the most common, with sulfur, phosphorus, zinc, and antimony being among the most popular. Some solids improve the performance of greases in severe applications, such as molybdenum disulfide, graphite, fluorocarbon powders, and zinc oxide. Polymers increase tackiness, low-temperature performance, and water resistance. The more popular polymers include polyisobutylene, methacrylate copolymers, ethylene-propylene copolymers, and polyethylene.

Reports of the effectiveness of grease are largely anecdotal. There are apparently too many indefinite variables involved for thorough analysis.

LUBRICATION WITH SOLIDS

Lubrication with liquids has both technological and economic limits. A technological limit is the physical and chemical degradation of a lubricant due mostly to temperature and acids, although such environments as vacuum, radiation, and weightlessness are also troublesome. In such cases, solid lubricants such as graphite or MoS_2 are used. Another limit of liquids is that chemically active (boundary) additives have not been found for such solids as platinum, aluminum, chromium, most polymers, and most ceramics. In such cases, a dispersion of solid "lubricant" in a liquid carrier may be applied. In other cases, such as in hot forming of steels, no additive is available for liquid lubricant; liquids evaporate and the low volatility hydrocarbons burn readily; and even if the liquid were to survive, its effectiveness would be very small at low speeds. In such cases, lime or ZnO may be a good (solid) lubricant, but these substances may be expensive to clean off in preparation for some later process. Also, liquid lubricants may be too expensive to use in some places. They require pumps, seals, and some way to cool the lubricant.

Solid lubricants in the form of graphite and MoS_2 were used in small amounts in the 1800s but research escalated from 1950 to 1965 when a wide range of loose powders, metals, oxides and molybdates, tungstates, and layer-lattice salts were investigated by the aerospace industry. Mixtures of graphite with soft oxides and salts in a variety of environments were also tried, as were coatings of silica in duplex structure ceramics and ceramic-bonded calcium fluoride. Overall it was found that solid lubricants should attach to one or both of a sliding pair to be effective for any reasonable length of time. Mica, for example, will not attach to steel and is ineffective as a lubricant; MoS_2 will not lubricate glass or titanium pairs, perhaps because these materials do not chemically react with the sulfur in the MoS_2.

Given the number of choices among available solid lubricants, it is apparent that logical and coherent classification of the types of solid lubrication is very difficult to achieve. However, solid lubricants may be functionally classified as shown in Table 7.2.

The effectiveness of a solid lubricant varies considerably with operating conditions, and it must be seen in the proper context. Solid lubricants of Groups A and B in Table 7.2 are often used where liquids are inadequate, and there is a finite possibility of part seizure (resulting in a shaft lockup or poor surface finish on rolled or drawn products). Thus, these lubricants are seen to be very effective

Table 7.2 Functional Groupings of Solid Lubricants

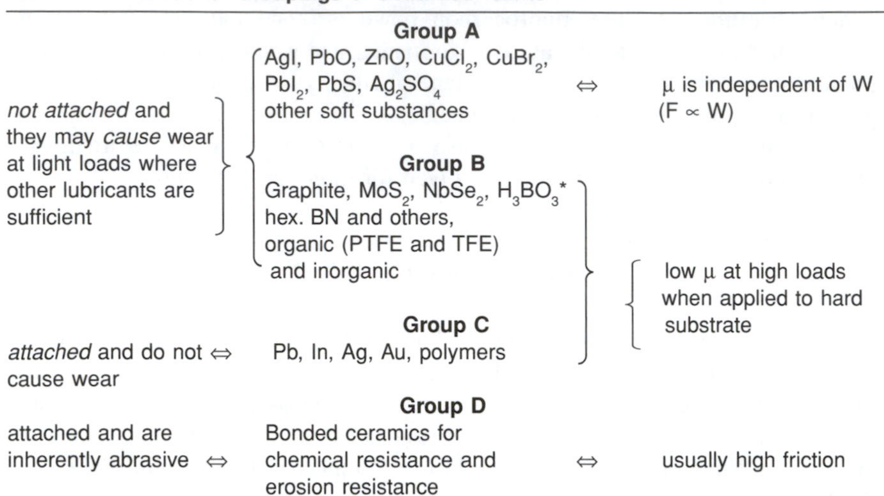

Attachment and wear		Substances		Friction behavior
		Group A		
not attached and they may *cause* wear at light loads where other lubricants are sufficient (Groups A and B)		AgI, PbO, ZnO, $CuCl_2$, $CuBr_2$, PbI_2, PbS, Ag_2SO_4 other soft substances	⇔	μ is independent of W (F ∝ W)
		Group B		
		Graphite, MoS_2, $NbSe_2$, H_3BO_3* hex. BN and others, organic (PTFE and TFE) and inorganic		low μ at high loads when applied to hard substrate (Groups B and C)
		Group C		
attached and do not cause wear	⇔	Pb, In, Ag, Au, polymers		
		Group D		
attached and are inherently abrasive	⇔	Bonded ceramics for chemical resistance and erosion resistance	⇔	usually high friction

* H_3BO_3 is boric acid in layered crystallite form which forms from B_2O_3 (a powder, which decomposes at ≈450°C) in moist air and functions up to 170°C. At 500°C it changes to boron trioxide. Graphite is a hexagonal structure, 1.42Å × 3.40Å spacing. MoS_2 is a hexagonal structure with S-Mo-S layers 6.2Å thick, spaced 3.66Å apart (covalent S-S bonds). Hexagonal BN has 2.5Å side dimension, layers 5.0Å apart, stacked in the order B-N-B.

in those cases. Unfortunately it becomes easy to expect benefit from these lubricants even where they are not needed. For example, if an engine oil is performing satisfactorily (i.e., there is some wear) anxious people add graphite or MoS_2 to the oil to reduce wear still more. Such products cost money, of course, in an amount that may exceed the savings due to prolonging engine life. At worst, even faster engine wear may be achieved at the higher cost! Solid lubricants are really abrasive to some extent, and they may wear engine bearing surfaces faster than dirt will or they might remove material faster than the loss of material by corrosion due to the additives in the oil. An example of the abrasiveness of a solid lubricant is the experience in an auto manufacturing company with the wear of bearings in the differential gear housing. It was found that some differential gear sets contained parts that had been marked with a grease pencil somewhere in the inspection sequence. These pencils contained ZnO, some of which fell into the lubricant and wore the bearings. This occurred even though the ZnO is thought to be softer than the bearings (>60 R_c) and in spite of the effectiveness of the EP additives usually found in differential gear oils. It was never resolved whether the ZnO removed boundary lubricant or whether it progressively removed the oxide from the steel.

Groups B and C in Table 7.2 provide low friction at high load. These substances (except Cr) function in the manner of the mechanism described by Tabor, where a "soft" surface layer has a low shear strength, but the surface layer is prevented from being indented by a hard substrate.

Graphite is the one of the three forms of carbon, and it functions as a lubricant. (Another form of carbon is diamond, the hardest substance on Earth and a covalent tetragonal structure. A third form is amorphous carbon.) Graphite, like MoS_2 is composed of sheets in hexagonal array, with strong bonding in the sheet and weak van der Waals bonding between sheets, providing low shear strength between sheets.

One major use of graphite has been as a brush material for collecting electrical current from generator commutators. Generators were used in airplanes until airplanes began to fly high enough to deprive the graphite brushes of air and water vapor. The brushes wore out so fast at high altitude that it was necessary to shorten high altitude flights. Oxygen and water vapor were found to be the most important gases. Bowden and Young[9] found the data sketched in Figure 7.8.

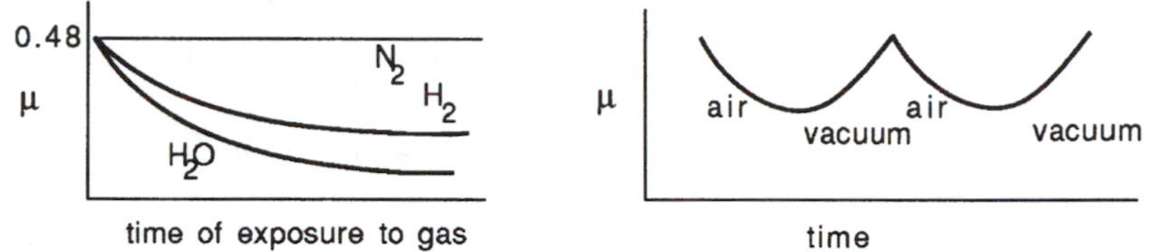

Figure 7.8 The influence of various atmospheres on the friction of graphite.

The effect of water vapor may be seen while peeling sheets of graphite apart in two environments. The work required to separate the sheets is expressed in terms of exchanging the interface energy of bonding between two sheets of graphite (γ_{GG}) for the surface energy of two new interfaces with vacuum (γ_{GV}):

$$\text{in vacuum } (2\gamma_{GV} - \gamma_{GG}) \approx 2500 \text{ ergs/cm}^2$$

$$\text{in water vapor } (2\gamma_{GV} - \gamma_{GG}) \approx 250 \text{ ergs/cm}^2$$

There is little effect of temperature even though one would expect that high temperature would drive off water.

MoS_2 works well in vacuum as well as in dry air. Water vapor affects MoS_2 adversely by producing sulfuric acid as follows:

$$2MoS_2 + 7O_2 + 2H_2O \Rightarrow 2MoO_2 \cdot SO_2 + 2H_2SO_4$$

(Sb_2O_3 inhibits corrosion in MoS_2 and improves gall resistance.)

Temperature affects the friction of both MoS_2 and graphite, as shown in Figure 7.9.

MoS_2 usually must be applied as a powder. It seems possible to electroplate the surface with Mo then treat with S-containing gas to obtain bonded MoS_2. However, bonding is most often best achieved with the use of carbonized corn syrup.

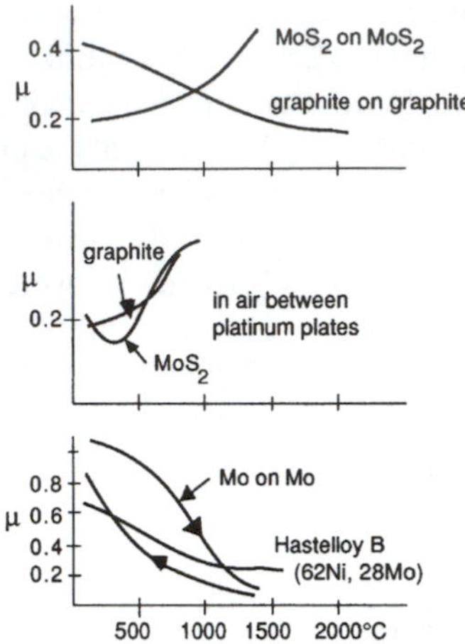

Figure 7.9 Friction of graphite and MoS_2 versus temperature.

Some practical advice on the use of solid lubricants was published by L.C. Kipp:[10]

1. All lamellars — keep liquids away, keep debris in, "sticky" substances work the best.
2. MoS_2 — limit to between 400°F and 700°F in air, and 1500°F in inert atmosphere
3. Limit PTFE to 550°F, FEP a little less.
4. Use graphite, in the range 400–1000°F, not in vacuum. Graphite causes galvanic corrosion because it is a conductor.
5. PbS and PbO are effective to 1000°F in air.
6. $NbSe_2$ is effective to 2000°F.
7. For bolt threads, burnish MoS_2 onto the threads up to 1000Å thick in an atmosphere without O_2 present.
8. CaF_2/BaF_2 eutectic, impregnated with nickel is effective from 900 to 1500°F.

REFERENCES

1. Cameron, A., *Principles of Lubrication*, John Wiley & Sons, New York, 1966.
2. Barwell, F.T., *Bearing Systems*, Oxford University Press, Oxford, U.K., 1979.
3. McKee S.A. and McKee, T.R., *Trans. ASME,* APM 57.15, 161, 1929.
4. Burwell, A.W., *Oiliness*, Alox Chemical Corp., Niagara Falls, 1935.
5. Cameron, A., Private communication.
6. Dowson D. and Higginson, G.R.G., *Engineering*, 192 158, 1961.
7. Wedeven, L.D., (Heard at a Gordon Conference.)
8. Musilli, T.G., *Lubrication Engineer*, May, 352, 1987.
9. Bowden F.P and Tabor, D., *Friction and Lubrication of Solids,* Oxford University Press, Oxford, U.K., 1954.
10. Kipp, L.C., *ASLE,* 32 11 574, 1976.

CHAPTER 8

Wear

SURFACES USUALLY WEAR BY TWO OR MORE PROCESSES SIMULTANEOUSLY. THE BALANCE OF THESE PROCESSES CAN CHANGE CONTINUOUSLY, WITH TIME AND DURING CHANGES IN DUTY CYCLE. WEAR RATES ARE CONTROLLED BY A BALANCE BETWEEN THE RATES OF WEAR, PARTICLE GENERATION, AND PARTICLE LOSS. PARTICLE GENERATION RATES ARE INFLUENCED BY MANY FACTORS INCLUDING THE NATURE AND AMOUNT OF RETAINED PARTICLES. THE LATTER IS STRONGLY INFLUENCED BY THE SHAPE OF A SLIDING PAIR, DUTY CYCLE, VIBRATION MODES, AND MANY MORE FACTORS. PRACTICAL WEAR RATE EQUATIONS ARE LIKELY TO BE VERY COMPLICATED.

INTRODUCTION

The range of wearing components and devices is endless, including animal teeth and joints, cams, piston rings, tires, roads, brakes, dirt seals, liquid seals, gas seals, belts, floors, shoes, fabrics, electrical contacts, discs and tapes, tape and CD reader heads, tractor tracks, cannon barrels, rolling mills, dies, sheet products, forgings, ore crushers, conveyors, nuclear machinery, home appliances, sleeve bearings, rolling element bearings, door hinges, zippers, drills, saws, razor blades, pump impellers, valve seats, pipe bends, stirring paddles, plastic molding screws and dies, and erasers.

Wear engages a major part of our technical effort. At times it seems that the rate of progress in the knowledge of wear is very slow, but while in 1920 automobiles could hardly maintain 40 mph for even short distances, they now go 80 mph for 1000 hours or so without much maintenance: this while adding greater flexibility, power, comfort, and efficiency.

The same is true of virtually every other existing product, although progress is difficult to perceive in some of them. We still have fabrics, television channel selectors, timers in dishwashers, and many other simple products that fail inordinately soon. Doubtless the short-lived products are made at low cost to maximize profits, but they could be made better if engineers put their minds to it.

Modern design activities are mostly evolutionary rather than revolutionary: most designers need only improve upon an existing product. The making of long-lived products requires considerable experience, however, not for lack of simple principles in friction and wear to use in the design process but because there are *too many* of them. The simpler notions still circulate, in design books and in the minds of many designers, such as:

1. Maintain low contact pressure
2. Maintain low sliding speed
3. Maintain smooth bearing surfaces
4. Prevent high temperature
5. Use hard materials
6. Insure a low coefficient of friction (μ)
7. Use a lubricant.

These conditions are not likely, however, to yield a competitive product. Designers need more useful methods of design, particularly computer-based methods. These are not yet available, certainly not in simple form as will be discussed more fully in Chapter 10.

In this chapter a perspective will be provided on what is known about various types of wear. Some machinery eventually fails or becomes uneconomical to operate because of single causes (types of wear), but most mechanical devices succumb to combinations of causes. A direct parallel is seen in the human machine. Medical books list various diseases, some of which are fatal by themselves, but usually we accumulate the consequences of several diseases and environmental contaminants along life's pathway. Predicting the wear life of machinery may perhaps be best understood in terms of the life expectancy of a baby. Both require the consideration of many variables and the interaction between them. In a baby these variables include family history, exposure to diseases and accidents, economic status, personal habits, social context of living, etc. Clearly, life expectancy is not a linear effect of the above variables, and the parallel breaks down in the determination of the endpoint of the process of decline.

One point of confusion in the literature on the subject of wear is the long list of terms that are used to describe types, rates, and modes of wear. The next section will list and define some of these.

TERMINOLOGY IN WEAR

One of the important elements in communication is agreement on the meaning of terms. The topic of wear has many terms, and several groups in professional societies have worked diligently to provide standard definitions for them. These efforts are largely attempts to describe complicated sequences of events (chemical, physical, topographical, etc.) in a few words, usually with minimal value judgment.

Following is a listing of 34 common terms used in the literature to describe wear. There are many more. Some terms communicate more than others the actual causes of loss (wear) of material from a surface, some are very subjective in nature and communicate only between people who have observed the particular wearing process together. Following are six categories of terms, progressing from the more subjective to the more basic. The latter terms are here referred to as,

MECHANISMS OF WEAR — *the* succession *of events whereby atoms, products of chemical conversion, fragments, et al., are induced to leave the system (perhaps after some circulation) and are identified in a manner that embodies or immediately suggests solutions.* These solutions may include choice of materials, choice of lubricants, choice of contact condition, choice of the manner of operation of the mechanical system, etc.

The grouping of terms:

1. The first group could be classified as *subjective or descriptive* terms in that they describe what *appears* to be happening in the vicinity of the wearing surfaces:

blasting	**hot gas corrosion**	**percussive**
deformation	**impact**	**pitting**
frictional	**mechanical**	**seizing**
hot	**mild**	**welding**

2. The second group contains terms that appear to have more meaning than those in group 1 in that some mechanisms are often implied when the terms are used. These types of wear do not necessarily involve loss of material but do involve some change in the sliding or contacting function of the machine.
   galling (may relate to surface roughening due to high local shear stress)
   scuffing } { probably relate to some stage of severe surface roughening
   scoring } { that appears suddenly in lubricated systems
3. **Adhesive** wear is the most difficult term to define. It may denote a particular type of material loss due to high local friction (which is often attributed to adhesion) and is a tempting term to use because high local friction produces tearing and fragmentation, whereas lubricants diminish tearing. Often **lubricated wear** is taken to be the opposite of adhesive wear.
4. Terms that derive from cyclic stressing, implying **fatigue** of materials:
   fretting, a small amplitude (few microns?) cyclic sliding that displaces surface substances (e.g., oxides) from microscopic contact regions and may induce failure into the substrate, sometimes generating debris from the substrate and/or cracks that propagate into the substrate)
   delamination describes a type of wear debris that develops by low cycle fatigue when surfaces are rubbed repeatedly by a small (often spherical) slider.
5. The fifth group can probably be placed in an orderly form but individual terms may not have originated with this intent. These relate to the types of wear known as **abrasive** wear. In general, abrasive wear consists of the scraping or cutting off of bits of a surface (oxides, coatings, substrate) by particles, edges, or other entities that are hard enough to produce more damage to another solid than to itself. Abrasive wear does not necessarily occur if substances are present that feel abrasive to the fingers! The abrasive processes may be described according to size scale as follows:

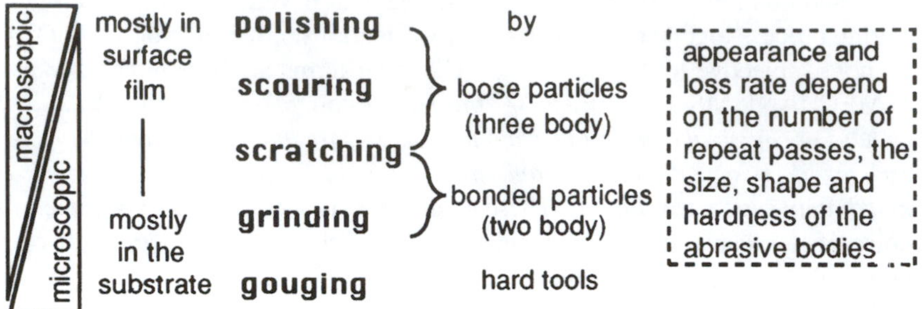

6. Wear by impingement, over angles ranging from near 0° (parallel flow) to 90°.

solid particle erosion
a. in a gaseous carrier
b. in a liquid carrier

liquid impingement erosion
causing **cavitation**

passing particles may cause **cutting**, particularly of ductile material

directly impinging particles are likely to cause **spalling** or brittle material and low cycle (plastic) fatigue of ductile material

HISTORY OF THOUGHT ON WEAR

Early authors on wear focused on the conditions under which materials wore faster or more slowly, but wrote very little on the causes of wear. In the 1930s the conviction grew that friction is due to an attractive force between solid bodies, rather than to the interference of asperities. The influence of this attractive force on friction became identified as the adhesion theory of friction, properly called a theory because the exact manner by which the attractive forces act to resist sliding was (and still is) not yet known. Some types of wear were also explained in terms of this same adhesive phenomenon, which led many authors to develop models of the events by which adhesion was responsible for material loss. Tabor described (in a word model) how dissimilar materials might fare in sliding contact where there is adhesion, as follows:[1]

Three obvious possibilities exist:

1. The interface is weaker (lower shear strength) than either metal — there is no metal transfer. An example is tin on steel.
2. The interface strength is intermediate, between that of the two metals, and shearing occurs in the soft metal. There is transfer of the softer material to the harder surface and some wear particles fall from the system. An example is lead on steel.

3. The interface strength is sometimes stronger than the hardest metal, there is much transfer from the soft metal to the hard metal, and some transfer of the hard metal to the soft surface. An example is copper on steel.

Not much can be said of these conditions because no one knows what the interface strength really is. Further, it should be noted that these examples generally describe transfer from one surface to the other, without stating *how* any of the transferred material is lost from the system as wear.

In the 1930s published papers began to distinguish between *adhesive* and *abrasive* wear:

1. Abrasive wear is thought by some to occur when substances that feel abrasive to the fingers are found in the system, and/or when scratches are found on the worn surface. Actually, scratches result from several mechanisms, and abrasive materials are abrasive only when their hardness approaches 1.3 times that of the surface being worn.
2. Adhesive wear was for many years thought to occur when no abrasive substances can be found and where there is tangential sliding of one clean surface over another. Oxides and adsorbed species are usually ignored. In 1953, J.F. Archard published an equation for the time rate of wear of material, Ψ, due to adhesion, in the form:[2]

$$\Psi = k\left(\frac{WV}{H}\right) = \left(\frac{N \times \frac{m}{s}}{Pa = \frac{N}{m^2}}\right) = \frac{m^3}{s} \tag{1}$$

where W is the applied load, H is the hardness of the sliding materials, V is the sliding speed, and k is a constant, referred to as a wear coefficient.

This equation is based on the same principles as Tabor's first equation on friction, discussed in Chapter 6, namely, that friction force, $F = A_r S_s$, where A_r is the real area of contact between asperities and S_s is the shear strength of the materials of which the asperities are composed. Archard assumed that $\Psi \propto A_r$ which in turn equals W/H for plastically deforming asperities, and $H \approx 3Y$ where Y is the yield strength of the asperity material. Each asperity bonding event has some probability of tearing out a fragment as a wear particle, which is expressed in "k," and the frequency of the production of a wear fragment is directly proportional to the sliding speed, V.

Archard's equation is one among hundreds of equations in the literature that are based on the phrase, "assume adhesion occurs at the points of asperity contact," or equivalent. Whereas adhesion is a reality, its operation between solids covered with the ever-present adsorbed species and wear particles is rarely examined, and no one shows how the presumed adhered fragments are released to leave the system as wear debris. However, Archard enjoyed the popularity of his model though he attributed it to "the sins of youth."[3]

In 1956, M.M. Kruschchov and M.A. Babichev published the results of a large testing program in abrasive wear. A curve fit to their data showed that:[4]

$$\Psi \propto \frac{WV}{H} \quad (2)$$

at least for simple microstructures. They, and later authors found more complicated behavior for other microstructures, which will be discussed in the section on *Abrasion and Abrasive Wear.*

The similarity in the above equations for abrasive and adhesive wear has been the source of confusion and amusement. Some authors concluded that since the wear rate is linearly dependent on either W, V, or H, or some combination, they must have seen abrasive wear predominantly. Others argued strongly for adhesive wear on the same grounds. The proponents of each mechanism have estimated what percentage of all practical wear is of their favorite kind, and the sum is much greater than 100%. *Further research is indicated!*

In the paragraphs that follow, there is no attempt to mediate between the proponents of abrasion and adhesion. Rather, some of the findings of careful research on the types of wear will be summarized.

MAIN FEATURES IN THE WEAR OF METALS, POLYMERS, AND CERAMICS

Dry Sliding of Metals

Let us consider wear during the dry sliding of clean metals. (Dry means no deliberate lubrication, and clean means no *obvious* oxide scale or greasy residue. Obvious means within the resolving capability of human senses. Recall that all reactive surfaces are quickly covered with oxides, adsorbed gases, and contaminants from the atmosphere.)

A. W. J. DeGee and J. H. Zaat[5] found that sliding produces two effects which are illustrated in Figure 8.1 for brass of various zinc content rubbing against tool steel. Brass is found to have transferred to steel where most of it remains attached, but some brass is removed (worn) from the system. The extent of each event depends on the Zn content in the brass.

1. Local adhering of brass to steel, for zinc content less than 10%. No iron is seen in the wear fragments. Some attached brass particles come loose from the steel but new material fills the impression again. Most of the steel surface remains undisturbed as seen by the unaltered surface features. The oxide on the brass is CuO. Possibly CuO + iron oxide *lubricates* well except at some few points, and at these points brass transfers to steel. (There was no analysis of possible oxide interphase.)
2. Continuous film, for zinc content more than 10%. The oxide on the brass is zinc oxide. Possibly this oxide does not *lubricate*. A thin film of brass is found

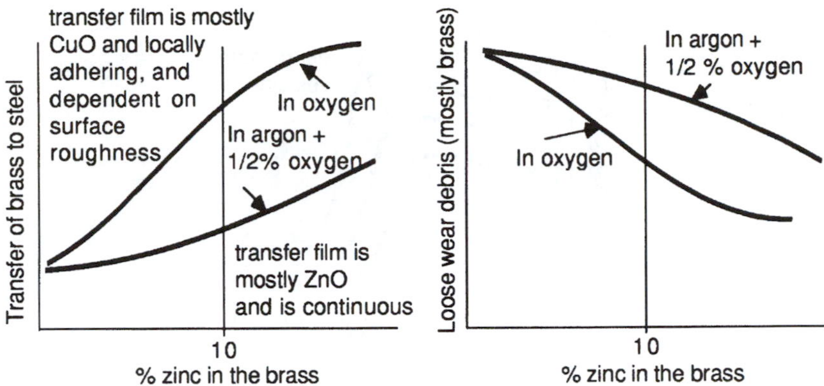

Figure 8.1 Variations in the rate of wear and rate of debris retention for brass.

on the steel. The wear particles are large but few. This film covers the surface roughness but wear continues. Thus this mechanism is not dependent on surface finish.

Lancaster[6] measured the wear rate of a 60Cu–40Zn brass pin on a high speed steel (HSS) ring over a very wide range of sliding speed and temperature, and got the results shown in Figures 8.2 and 8.3. He classified wear in relative terms, mild and severe — severe in the region of the peaks of the curves and mild elsewhere.

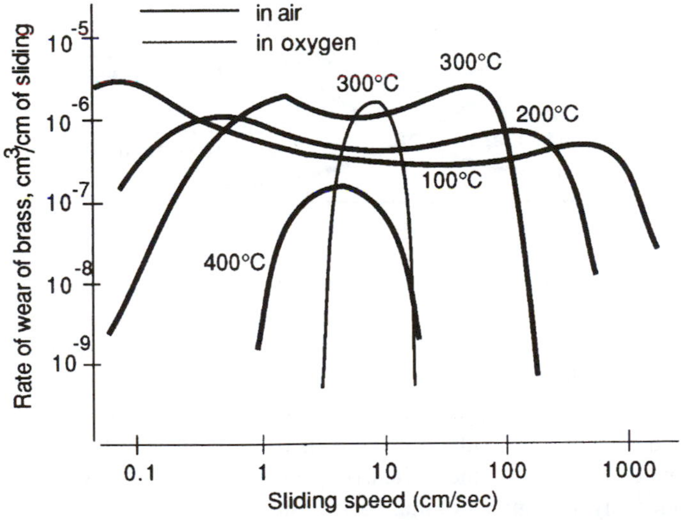

Figure 8.2 Wear rate versus sliding speed, with 3 Kg load.

The transition between severe wear and mild wear is influenced by atmosphere, as well as sliding speed and ambient temperature. Figure 8.3 suggests

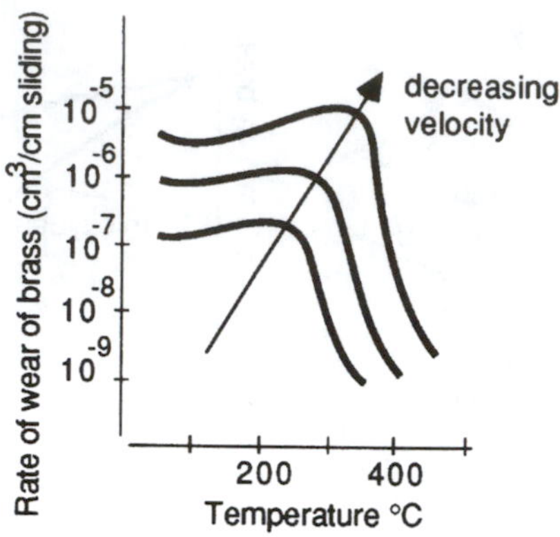

Figure 8.3 Wear rate versus temperature.

that sliding causes sufficient surface heating to offset some of the effects of ambient heating. Note the influence of atmosphere.

Lancaster proposed that the transition between mild and severe wear was influenced by the thickness of oxide. The oxide thickness is a function of two factors, namely, the time available to reoxidize a denuded region (on the steel ring) and by the rate of formation of the oxide as sketched in Figure 8.4. The time available to oxidize is determined by sliding speed in repeat-pass sliding as with a pin on a ring. The rate of formation is influenced by temperature rise due to sliding at the denuded region as well as by the ambient temperature.

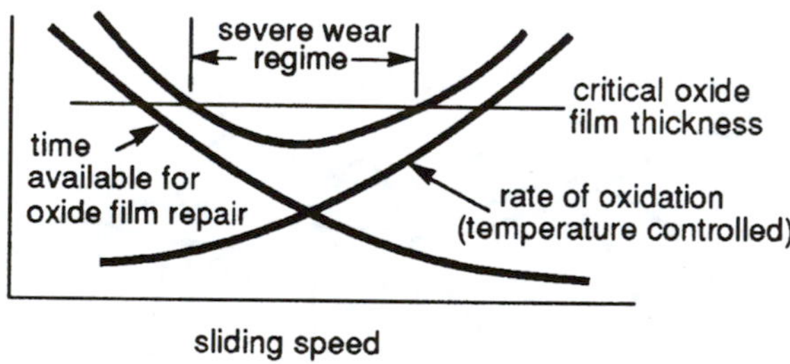

Figure 8.4 Influence of competing factors that control oxide film thickness.

Figure 8.5 compares the wear rate of the steel ring with that of the brass pin. The different locations of the transitions of the two metals are probably as much related to metal and oxide properties as to the geometry of the specimens.

Figure 8.6 shows the result of an analysis of the surface of the brass pin, after sliding, to a depth of 0.005 inch. Clearly, the brass does not slide directly on the steel but on a layer of mixed oxide, metal, and adsorbed substances.

Finally, Figure 8.7 shows the relation between wear rate (ψ), the coefficient of friction (μ), and electrical contact resistance over a range of temperature. Apparently at the higher temperatures there is sufficient oxide to electrically separate the metals, and to increase μ.

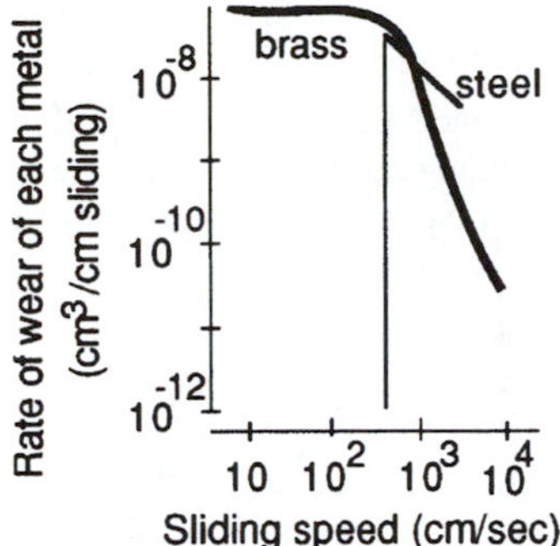

Figure 8.5 Wear rates of brass pin and steel ring.

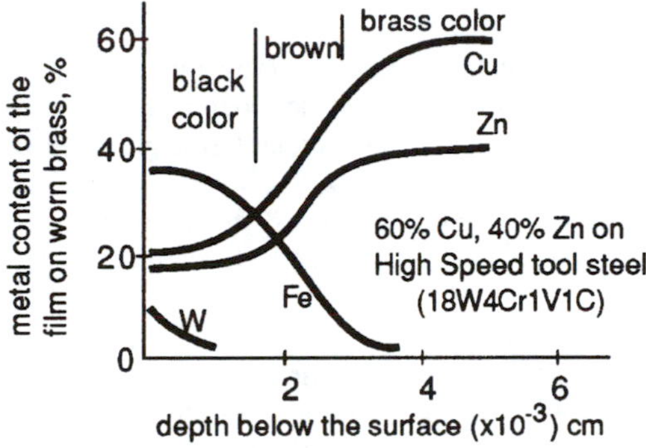

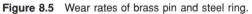

Figure 8.6 Surface composition of worn brass (60–40) pin.

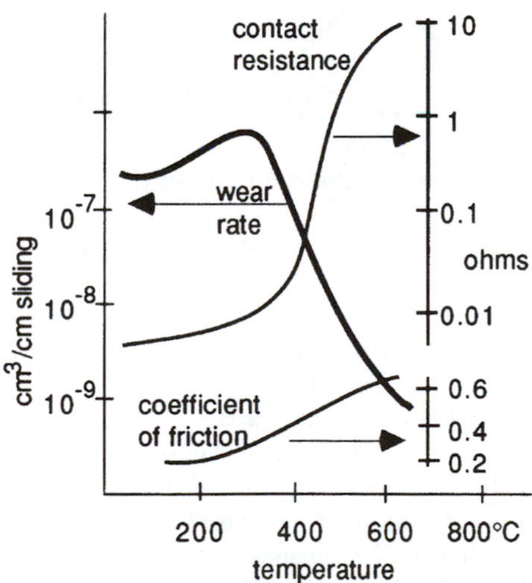

Figure 8.7 Friction and wear.

N. C. Welsh[7] worked with two steels, 0.12% and 0.5% carbon steels:

1. The 0.12% carbon steel: increasing the applied load may decrease wear rate as shown in Figure 8.8. Sliding raises the temperature in the contact region, and the higher load may heat the steel into the austenite range.

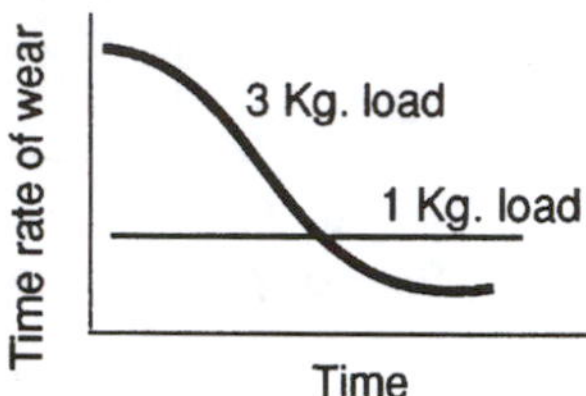

Figure 8.8 Wear rate versus time for two loads, low carbon steel.

Apparently, nitrogen from the atmosphere (and carbon from a lubricant) dissolves into the austenite. The metal then cools quickly and the former pearlite grains become martensite, and some former ferrite grains become strengthened by nitrogen. The net effect is to lower the wear rate after many local regions (asperity dimensions) become hardened. Partial proof of the surface hardening mechanism may be seen in Figure 8.9, which compares steels of high and low hardenability.

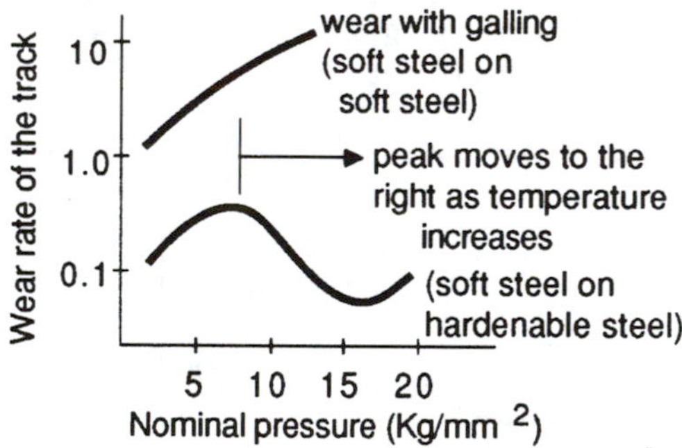

Figure 8.9 Comparison of wear rates of unhardenable versus hardenable steels.

Figure 8.10 suggests, however, that oxidation is also important, and may be influenced by hardness: the contact pressure at which wear rate is high coincides with high metal content in the debris.

2. Welsh later measured ψ versus load for 0.5%C steel on steel, using a pin-on-ring configuration and found transitions between severe wear and mild wear.[8] His data were published in the form shown in Figure 8.11, from which three curves were selected for illustrative purposes.

   The large transitions ($\approx$ 2.5 orders of ten) in the data for the softest steel seem impossible and yet they are real: these data for 1050 steel as well as for other steels have been verified by research students many times.

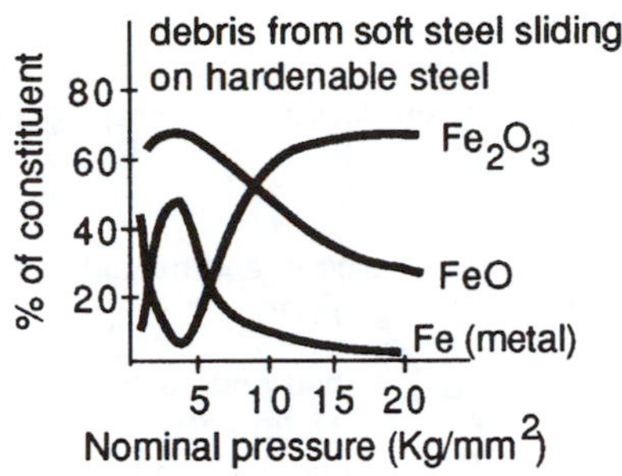

Figure 8.10 Composition of wear debris in tests of Figure 8.9.

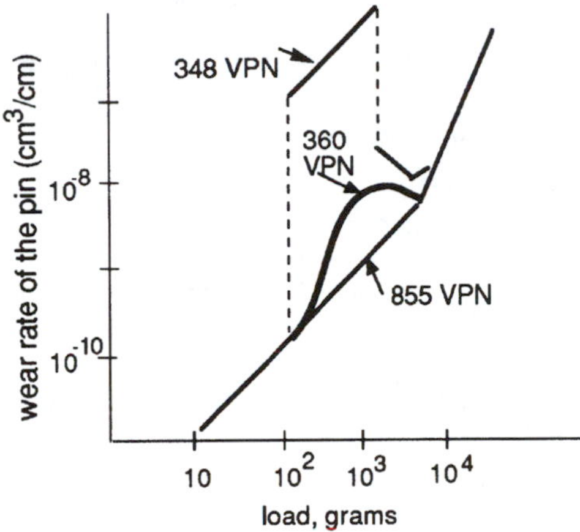

Figure 8.11 Wear rate versus load for 1050 steels of three hardnesses. (Adapted from Welsh, N.C., *Phil. Trans. Roy. Soc.* (London), part 2, 257A, 51, 1965.)

The effect of hardness is to diminish the extent of transition to severe wear. It may be speculated that the critical oxide thickness is less for hard substrates than for the soft substrate.

Additional effects were noted. For example, sliding speed influenced the transitions and so did atmosphere, as shown in the sketch below, showing the effect on the upper sloping line in Figure 8.11.

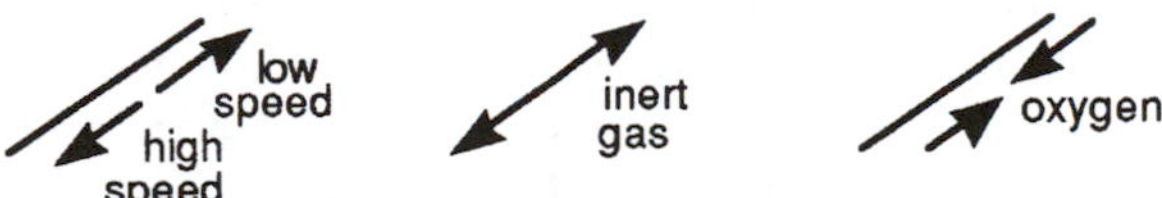

Figure 8.12 shows the accumulated weight loss of the ring in Welsh's experiments. In the mild wear regime, initial ψ was high at the first sliding of newly made surfaces and after oxide is removed chemically and rubbing resumes. Welsh explained this in nearly the same terms as did Lancaster, as illustrated

in Figure 8.13. The apparent lower sensitivity of the 855 VPN hard steel to load in Figure 8.11 may be due to greater effectiveness of thin films of oxide on hard substrates than on soft substrates. Perhaps oxide films *do* lubricate materials.

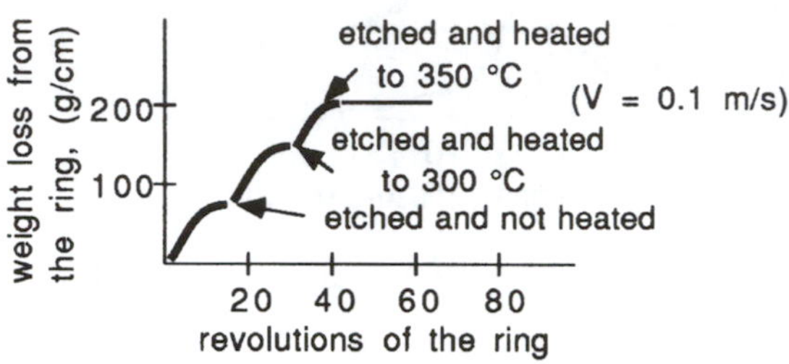

Figure 8.12 Effect of heating steel on wear rate.

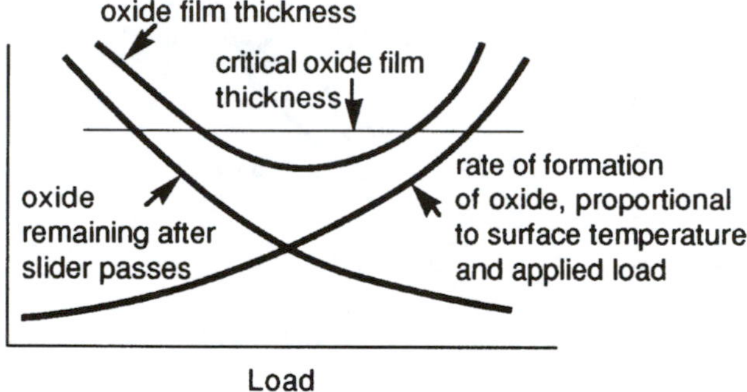

Figure 8.13 Schematic representation of two factors that may influence the thickness of oxide coatings, as a function of applied load.

What mathematical expression can be formed from the above data? Does Archard's equation (Equation 1) suffice? ***(Also, see Problem Set questions 8 a, b, c, and d.)***

Oxidative Wear

The discussion above shows that the oxides of metals prevent seizure (galling, adhesion) of metals together. (Seizure, galling, etc., are likely to occur in vacuum where oxides grow slowly, if at all.) In the common condition of sliding when oxides are prominent, wear certainly occurs, but there is some confusion in the literature as to how to categorize this type of wear. In early years, it was described as abrasive because it clearly was not adhesive. As will be discussed below, the designation "abrasive wear" is not satisfying either, because abrasion is defined in terms of the presence of hard substances in the interface region. When oxide particles are loosened and move about within the contact region, they loosen more particles, some of which leave the system as wear debris, but the oxides do not abrade the substrate in most systems. Wear by loosening of and loss of oxide should therefore not be identified as abrasive wear.

The rate of formation of the oxides is the basis for the oxidative mechanism of wear formulated by Quinn[9] in the following equation:

$$\omega = \frac{WdA_p e^{\left(\frac{-Q}{RT_o}\right)}}{Up_m f^2 \rho_o^2 \xi_c^2} \tag{3}$$

where ω is the wear rate per unit distance of sliding, W is the applied load, d is the distance of sliding over which two particular asperities are in contact, U is the sliding speed, p_m is the hardness of the metal immediately beneath the oxide, f is the fraction of oxide which is oxygen, ρ_o is the density of the oxide, ξ_c is the critical thickness at which the surface oxide film becomes mechanically unstable and is spontaneously removed to form the basis of the wear process. A_p and Q are oxidational parameters, R is the gas constant, and T_o is the temperature at which the surfaces of the sliding interface oxidize.

The mechanism of wear envisioned by Quinn is that a sliding surface heats up and oxidizes at a rate that decreases with increasing oxide film thickness. At some point the film reaches a critical thickness and flakes off. Thus the thicker the film (larger ξ_c) becomes before it separates, the more slowly oxides form overall and the slower will be the wear rate.

Quinn's equation has been frequently discussed but it is not an adequate description of the coming and going of oxide. His theory offers no role for friction stresses in the removal of oxide, but rather is based on spontaneous loss of oxide when it reaches a particular thickness. Further, Quinn focused on very thick oxides, such as furnace scale, which is very different from the oxide on most surfaces.

Following is a short discussion that has become common knowledge among tribologists. It describes oxides of iron, formed in air, without sliding:

> Iron forms three stable oxides, wustite (Fe_xO), where x ranges from 0.91 to 0.98, magnetite (Fe_3O_4, opaque, SG≈5.20, MP≈1594°C), and hematite (Fe_2O_3, transparent, SG≈5.25, MP≈1565°C). The Fe_xO has less than a stoichiometric amount of Fe (rather than an excess of O_2) and has the NaCl type of cubic structure. It is a "p" type (metal deficient) semiconductor in which electrons transfer readily. Fe_3O_4 seems also to be slightly deficient in Fe but is regarded as having an excess of O_2. Its structure is (spinel) cubic. There are three structures of Fe_2O_3, namely, alpha which has the (rhombohedral) hexagonal structure, beta which is uncommon, and gamma which has the cubic structure much like Fe_3O_4. Fe_2O_3 is an "n" type (metal excess) semiconductor, in which vacancy travel predominates.
>
> The type of oxide that forms on iron depends on the temperature and partial pressure of O_2. At temperatures above 570°C, first O_2 is absorbed in iron solid solution, then Fe_xO forms, which in turn is covered with Fe_3O_4, and then Fe_2O_3 as the diffusion path for Fe++ ions increases. Below 570°C there forms, simultaneously, a thin film of FeO (MP≈1369°C) under a film of Fe_3O_4.

Fe_xO and Fe_3O_4 can be oxidized to the more O_2-rich forms of oxide, and H_2 or CO can reduce Fe_3O_4 and Fe_2O_3 to lower forms of oxide and can reduce Fe_xO to elemental iron.

The rate of oxidation of iron and steels is nearly logarithmic. At room temperature the oxides of iron asymptotically approach 25Å in 50 hours. These rates can be altered by alloying. An "n" type oxide can be made to grow more slowly by adding higher valency alloys than that of the base metal, and vice versa.

In moist air, FeOH (it is green or white with SG≈3.4) may form, or even $Fe_2O_3 \cdot H_2O$ (red/brown powder, SG≈2.44–3.60).

Dry Sliding Wear of Polymers[10]*

Plastics: The friction of plastics is about the same as that of metals, except for PTFE (at low sliding speed only), but the seizure resistance of plastics is superior to that of soft metals. There is general uncertainty about the influence of surface roughness on wear rate, and some polymers wear metals away, without the presence of abrasives.

The general state of understanding of polymer wear is that rubbing surfaces experience a break-in period, followed by a steady wear behavior, often referred to as linear wear. It is in the linear region that most people have been searching for useful wear coefficients.

A second quantity is some descriptor of the rubbing severity above which severe or catastrophic wear may occur. The most widely known descriptor is the PV limit, where P is the average contact pressure (psi) and V is the sliding speed (fpm). Each polymer has a unique PV limit as measured by some test, most often a "washer" test. It is apparently a thermal criterion taken from the idea that PV, multiplied by the coefficient of friction, μ, constitutes the energy input into the sliding interface. (See equations in the section titled *Surface Temperatures in Sliding Contact* in Chapter 5.) If the energy is not removed at a high enough rate, the polymer surface will reach a temperature at which it will either melt or char, and severe wear will occur. There are three compelling reasons for doubting this hypothesis. The first reason is that there is not as sharp a decrease in μ when severe wear occurs as one might expect if molten species were to suddenly appear in the contact region. The second reason is that the published PV limits are not in the same order as the melting points for a group of polymers. For example, the limiting PV at 100 fpm for unmodified acetal (MP≈171°C) is 3000 and that for Teflon® (MP≈327°C) is 1800. The third reason is that for some polymers, gas is evolved from the region of sliding when operating in the regime of "mild wear," and these gases are known to form at temperatures well above the melting point of the polymer.

The nature of transfer films is important in the wear process. Some films, as from pure PTFE and polyethylene, are smooth and thin, as thin as 0.5μm, are not visible, and must be viewed by interference methods. Other polymers produce thick, discontinuous, and blotchy films. If a film of polymer is formed on the metal counterface and it remains firmly attached, the loss of the polymer from the system is minimal after the first pass in multipass sliding, and mysteriously, the friction

* The first synthesis of polymers occurred in 1909.

often decreases as well. If during sliding a particle of polymer is removed from the polymer bulk but does not remain attached to the metal, it is lost from the system. An intermediate state of wear is the case where a transfer film is formed, but fragments of the film are later lost, probably due to fatigue or some other mechanisms. These fragments, or wear particles, may be very small: fragments from the ultra-high-molecular-weight polyethylene (UHMWPE) in prosthetic hip joints are small enough (<1μm) to find their way into distant body organs. Since the behavior of the system is very different in three regimes, the discussion will now focus in turn upon some aspects of the break-in period, the steady wear regime, and the severe wear regime. Most such research has been done under laboratory conditions. Practical conditions are more revealing but usually sparsely documented. In practice most polymers are exposed to lubricants and other conditions that dissolve into and alter surface properties. For example, paraffinic hydrocarbons react with rubbers and ethylene-propylene; ethers and esters react with polycarbonates, polysulfones, and rubber; and silicone liquids react with silicone solids.

In the early stages of sliding the rate of buildup of transfer film is dependent on the orientation of surface finish relative to the sliding direction, and varies with the type of contaminant or dirt on the sliding surfaces. Surface finish appears to have almost no effect on the steady-state wear rate, probably because the products of wear fill the grooves in the surface.

To check the effect of surface roughness on wear in the early stages of sliding, several polymers were slid on carbon steel with surface roughnesses from 0.1 μm to 3 μm Rq, some parallel with the sliding direction and some perpendicular. For Nylon 6-6 at a speed of 0.4 m/s the data in Figure 8.14 are obtained. It may be seen that nonlinear or break-in time may persist twice as long with parallel sliding as with perpendicular sliding and that the weight loss at the end of break-in may vary by a factor of 4 or more. The break-in period is a time when a film of polymer is transferred to the metal. The equilibrium film thickness for all tests run on various surfaces at one speed and one load were about the same.

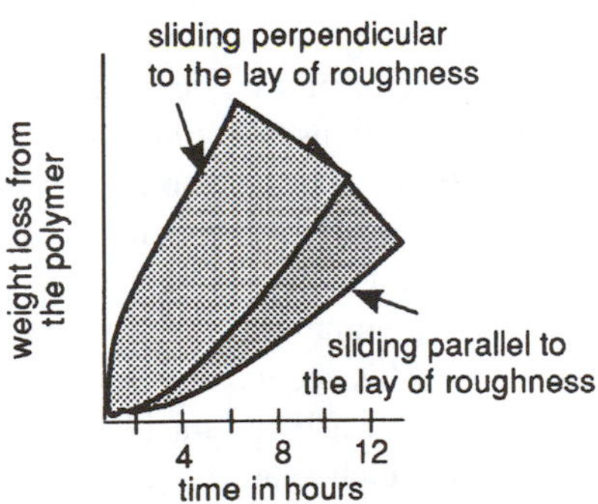

Figure 8.14 Influence of surface roughness on wear rate of a polymer.

The second factor in controlling the establishment of the transfer film is surface cleanliness. Tests were done with metal surfaces in three conditions: namely, laboratory clean (an adsorbed water film), a thin film of inert hydrocarbon (vacuum pump oil), and a machine oil. In all cases transfer films begin to form and they may become continuous, each at a different time. The higher the temperature of the countersurface, the more quickly the transfer film forms.

(See Problem Set questions 8 e and f.)

As the transfer film forms, the loss rate from the polymer is high, but after the film is formed the wear rate is much lower, sometimes less than 1% of the initial rate. Thus, predicting total wear rate of a bearing over some specified time should not be done on the basis of the steady-state wear rate alone. For short time use of a polymer bearing, the break-in stage could produce much more wear than an amount based on predictions from data for steady-state wear. Further, wear rate predictions are complicated by variations in temperature, variations in amounts of contamination, by variations in speed, by start-stop or forward-reverse cycles, and other factors.

Research was done to determine the validity of the temperature criterion for the onset of severe wear. Pins of several polymers (PE, POM, Nylon 6-6, Delrin AF) were rubbed against 440C stainless steel in a vacuum. A thermocouple was embedded in the steel and a magnetic-sector-gas-analyzer was placed into the vacuum chamber. The latter provided information on the gases emitted from the sliding interface, in terms of the ratio (atomic-mass-units/electron charge). To calibrate the latter, small bits of polymer were heated to various temperatures, and profiles were obtained of the emitted gases. When these same profiles were seen in the sliding experiments, the surface temperature in the interface was known.

The temperatures as measured by the thermocouple and by the gas analyzer did not correspond well. From these tests it was found that even though the sliding surface temperature was appreciably higher than the crystalline melting point and the softening point of the polymer, and actually reached the thermal degradation temperature, no measurable wear occurred until the transfer film was removed, which occurred when the steel surface reached a temperature in excess of 50°C above the softening point of the polymer tested. Severe wearing occurred at that point.

The transition to severe wear occurs by the following sequence of events. During low-wear-rate sliding conditions, the transfer film remains as a flat film behind the slider and provides (or becomes) a lubricant film upon which the slider rides on later passes. If the temperature of the transfer film is high, such that the film of polymer has low viscosity, and if the low viscosity polymer does not wet the metal surface, the polymer agglomerates into spheres which are removed by the next slider which passes by. This sequence is seen in Figure 8.15.

Thus the sliders are deprived of a lubricant film, and instead give up some material to establish a new film which again is quickly detached. The difference between a tenacious and fleeting transfer film is a wear rate that varies by more than 2000 times. Relevant variables in this entire process must include the temperature and thermal properties (mass, thermal diffusivity, etc.) of the countersurface, and contact area between the sliding parts.

The overall effect of these mechanisms of material transfer and loss is a wear rate that may be sketched as shown in Figure 8.16. The wear rate increases with

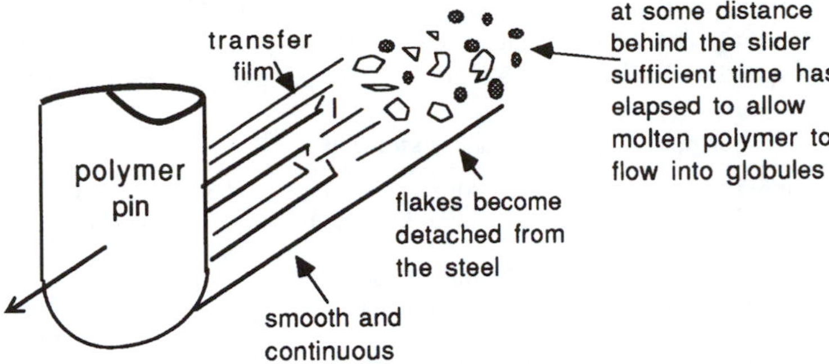

Figure 8.15 Wear mode of polymer pin sliding on flat.

sliding severity, which is some combination of P, V, and other variables. That wear rate is most likely due to some combination of inadequate attachment of transfer film to the substrate and removal of transfer film by fatiguing and other failure modes. If the surfaces are clean and attachment strength is high, at high severity of sliding the transfer film does not readily fragment by fatigue, and wear is low. At high severity the transfer film, however firmly or weakly attached, agglomerates and is lost as wear debris.

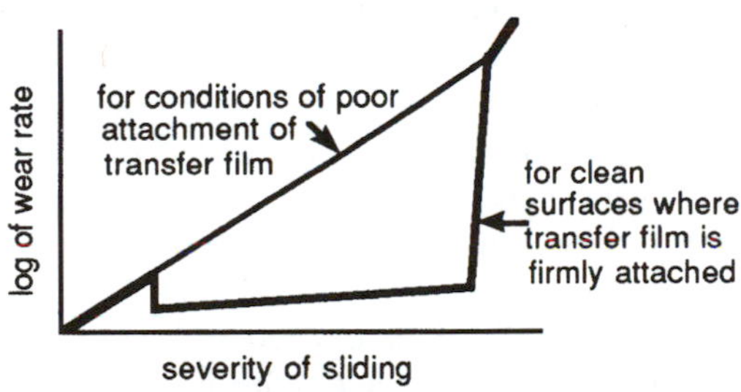

Figure 8.16 Wear rate for two surface conditions.

Metal wear by plastics: When sliding some of the harder plastics on 440C stainless steel, hardened to 50 Rc, at all speeds and loads, Fe and Cr were found attached to the polymer at the end of the test in the mild wear regime. These results indicate, first, that a lamellar transfer film is not laid down by successive and simple shear from the polymeric sliding. Rather, there is considerable turbulence or rolling of polymer within the transfer film, at least in the early life of the transfer film.

Second, the transfer of Fe and Cr to the polymer indicates that the metal is wearing away. This was verified by sliding for 50 hours, after which the amount of wear could be measured by a (roughness) tracer profile of the sliding track. To make sure that there were no abrasive substances in the polymers, several were obtained in which no vanadium had been used as a catalyst in the polymer-making processes. Thus there was no hard vanadium oxide in the polymer. These polymers also wore metal away.

Soft plastics did not wear away the 440C steel of 50 Rc hardness (TS≈1.25 GPa). The list of plastics used in these experiments is given below for a sliding speed of 1.47 m/s and a load of 222N per pin of 13mm diameter:

Polymer	Shear strength	μ	
Nylon 6-6	70.5 MPa	0.66	these wear steel away
Delrin (POM)	65.5 MPa (9500 psi)	0.65	
HDPE	63.4 MPa (9200 psi)		these do not wear the steel
Delrin AF (POM + PTFE)	55.2 MPa (composite)	0.20	
Nylon 11	41.4 MPa	0.50	
UHMWPE	24.1 MPa	0.55	

Rubber is a polymer but it differs from the plastics in that its molecules are crosslinked. Thus the migration or flow of the molecules is severely limited.

The English chemist Joseph Priestley gave rubber its English name in 1770, because the new substance in his hand would rub out pencil marks. (Rubber is known as elastomer in some languages.) Enough rubber is produced each year to cover the city of Ann Arbor, Michigan (or the region within the Circle underground line in London) with a 19mm blanket. That is about 10^{10} kg. Forty percent of that production is natural rubber, and the remainder is made up of about eight synthetic types. About 20% of this volume of rubber is worn away, and another 43% is discarded because of the volume that is worn away, mainly in tires.

Rubber wears by two mechanisms, tearing and fatigue.[11] Ultimately, these are not very distinct mechanisms, because tearing, or fracture, is failure in $^1/_4$ cycle of fatigue. In general, both forms of failure arise from high local friction against the opposing surface (rough particles or smooth surfaces) relative to the strength of the rubber. The friction is stated to be local friction (asperity scale) rather than measured macroscopic friction: severe wearing can often occur though the measured friction is low (or at least not high). For several types of rubber the wear rate increases by orders of 10 when μ increases, e.g., from 1.0 to 1.2.

Wear of rubber by virtually all causes is referred to as abrasion even though the opposing surface may not appear to be abrasive. Wearing does not appear to result from progressive removal of chemically altered surface material, but rather by removal of chemically unaltered molecular chains. The size scale of wear fragments ranges from fractions of μm to mm: the smaller dimension producing surfaces that appear shiny, the latter, matte.

The tearing mechanism is immediately visible. It occurs when sliding on rough surfaces, particularly on a surface of sharp stones (for tires) or abrasive paper. This conclusion is supported by the observation that the ranking of several rubbers in a test on abrasive surfaces is the same as in tensile tests of the rubber. Even the temperature dependency and rate dependency are the same.

The fatigue mechanism occurs when rubber slides on undulating surfaces without sharp protrusions. This mechanism is supported by a correlation between the distance of sliding (number of deformation cycles due to passing bumps) until surface failure occurs and the number of strain cycles to tensile failure. It is supported by a parabolic relationship between applied stress and fatigue life in both tests. The relationship is further confirmed by noting that oxygen in the atmosphere increases wear rate and antioxidant in the rubber decreases the wear rate.

In the literature on the wear of rubber the term "pattern abrasion" is often seen. The term refers to the texture seen on worn surfaces. Particularly in the fatigue mode of wear the rubber is fractured at regular intervals with the fracture extending downward at an angle of about 15° from the surface in the direction of sliding. These fracture planes are joined by a cross fracture, also at fairly regular intervals. The direction of sliding may be determined from the pattern. Most of the pattern marks are perpendicular to the sliding direction in dry sliding, but there are likely also to be some marks in the direction of sliding when abrasives are present, particularly in lubricated sliding.

Carbon-black filled rubber, such as tire rubber, is much stiffer than unfilled rubber and produces much lower friction against other surfaces than does unfilled rubber. However, it also has less ductility, but greater damping loss, usually. The balance of these properties strongly influences wear rate, but the optimum balance depends as strongly on the mechanical structure *holding* the rubbing component. For example, in addition to the microscopic stress fields in sliding surfaces the macroscopic shear stress that is imposed upon the tire–road interface in braking is higher toward the rear of the contact patch than toward the front. The rubber is passing through the varying strain field. By contrast, experiments on the wearing of rubber are often done with blocks of rubber that slide over their entire surface at once. The rubber in the sliding block has a constant macroscopic state of stress imposed upon it.

Wear of Ceramic Materials[12]

General Features of Wear: There are four fairly consistent differences between metals and ceramic materials in sliding contact:

1. The coefficient of friction of ceramic materials is usually significantly higher than that of metals. A parallel behavior is that ceramic materials are much more likely to produce severe vibrations during sliding than do metals.
2. In repeat-pass sliding with the pin-on-disk specimen shape, the wear loss from the pin is greatest for metal combinations (unless the disk is much softer than the pin), whereas the wear loss from the disk is greatest for ceramic combinations. There is often little wear in the early stages of sliding, followed in time by a rising wear rate.
3. The wear rate often increases sharply at some point during an increase in sliding speed, probably due to thermal stress cycling.
4. The wear rate often increases sharply at some point during an increase in contact pressure. An explanation is given below under *Wear Models for Ceramic Materials*.

Ceramic materials are different from metals and polymers in two very important respects that influence wear and surface damage:

1. The grains are brittle (but do behave in a somewhat ductile manner under compressive stress). Ceramic materials are mostly either ionic or covalent structures. Thus there is an overall brittle behavior of macroscopic-size specimens in a tensile test or impact test.
2. Grain *boundaries* range in properties from very ductile to very brittle. The reason is that many generic ceramic materials are made with several different, often ductile, sintering aids that become thin second phases or intergranular (grain boundary) materials. Si_3N_4 often has MgO or Y_2O_3 grain boundaries. SiC usually has none and in such materials the anisotropic behavior of grains places a high stress on grain boundaries during temperature changes and with externally applied stresses. ZrO_2 is an example of ceramic material that changes lattice structure under stress, from the tetragonal phase to the ≈5% less dense monoclinic phase under tension, reverting partially to the tetragonal phase under compression.

These distinct properties produce two effects in tribological applications that are less obvious in other applications:

1. The small scale nonhomogeneous strain fields induced in materials in sliding or erosion preferentially fracture brittle grain boundaries.
2. The anisotropic morphology of ceramic materials promotes failure in repeat stress applications, also known as fatigue behavior. Since many tribological situations involve repeat-pass sliding, repeat impacts, etc., a fatigue mode of ceramic wear may be prominent. In the ceramic materials with ductile grain boundaries, the fatigue mechanisms are similar to the low-cycle fatigue mechanisms in metals. In the ceramic materials with brittle grain boundaries, failure also occurs in few cycles but cracks propagate quickly because of high residual and anisotropically induced stresses.

Wear Models for Ceramic Materials: The most formal thinking on wear mechanisms of ceramic materials focuses on their brittle behavior. Wear is assumed in many papers to occur by the damage mechanisms formed by a sharp static indenter. Cracks occur at the corners of indentations made when a load is applied upon a Vickers or Knoop indenter, producing planar cracks perpendicular to the surface. Cracks also appear at some depth below the surface when the load is removed from the indenter. These are oriented parallel to the surface and are the result of plastic flow during indentation. A sketch of these cracks is shown in Figure 2.20 in Chapter 2.

Several equations have been derived using the principles of indentation fracture mechanics (IFM). The most widely discussed is the work of Evans and Marshall.[13] They assume that material removal begins as a loosening of material by linking of the two types of cracks that develop under a sharp indenter. A sharp slider extends the crack system over a distance, S, to produce a wear volume, ς:

$$\varsigma = \frac{\left(W_n^{\frac{9}{8}}\left(\frac{E}{H}\right)^{\frac{4}{5}} S\right)}{K_c^{\frac{1}{2}} H^{\frac{2}{5}}} \quad \textbf{(4)}$$

where

W_n = normal contact force
K_c = fracture toughness
E = elastic modulus
H = hardness
S = sliding distance

Evans and Marshall reported a qualitative correlation of this equation with the wear rate of glass, but very poor correlation with polycrystalline structural ceramics. It is interesting that most authors quoting the above work overlooked these latter poor results. In fact, a few others have confirmed these hypotheses, but usually not over a wide range of test variables.

The implication in the work of Evans and Marshall is that damage due to sliding occurs on a large or macroscopic scale, mostly because the crack patterns that provide the basis for their hypothesis were made by large-scale indenters compared with the grain sizes of ceramic material (which are generally in the range of 1 to 10 μm in diameter). Actual wear debris particles are usually not macroscopic in size.

Microscopic Wear Damage: Ajayi[12] found that wear rates (of four very different materials) could not be correlated with materials properties as suggested in the IFM approach. Further, the wear debris was microscopic in size, that is, very much smaller than the apparent contact diameter between his slider and flat disk. It was not possible to determine whether the wear debris began as microscopic particles or whether it began as larger particles and was crushed in later passes of the slider. Ajayi used spherical sliders and considered whether his differences from the hypotheses of the IFM approach may have been due to slider shape. He therefore indented flat surfaces with both sharp and spherical indenters. In both cases he found that fragmentation of the plate materials occurred within the contact area.

Lee[14] repeated some of Ajayi's experiments, applying a higher range of loads, with synchronized vertical and horizontal cyclic load on a sphere and flat plate. He also found that fragmentation occurred on a microscopic (grain size) scale, which progressed with the number of load cycles and progressed at a much greater rate when high tangential force was applied. His data suggest a strong fatigue effect and a significant sensitivity of wear rate to the coefficient of friction. The latter is usually not controlled in a wear test.

Since Lee worked with a spherical indenter he did not find cracks radiating from the indentation. Rather, at very high loads, ring cracks appeared around the indentation, again accelerated by cyclic stressing. The normal load required to produce ring cracks was generally at the highest end of the range of loading in the separate tests of Ajayi and Lee. (At least five equations are available in the literature for estimating the loads required to cause ring cracks. Each uses a different assumption of initiating crack length or fracture energy, resulting in a wide range of estimates.)

The scale of microdamage is compared with the size scale of ring cracks in Figure 8.17. Note in the sketch that the edges of the ring cracks also fragment, providing a second source of wear debris. It is this second source that probably causes the great increase in wear rate when high contact stresses are imposed on ceramic materials.

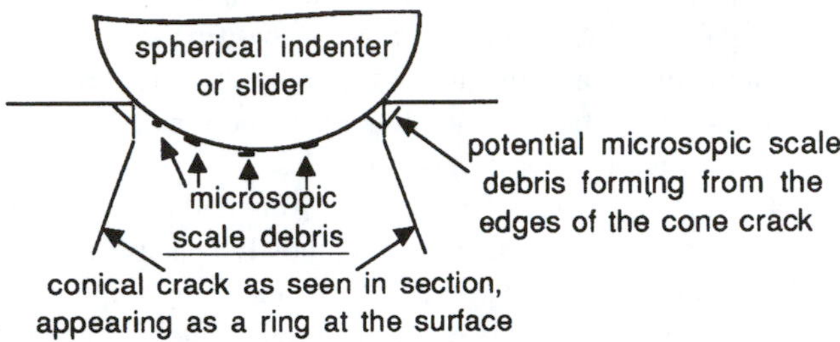

Figure 8.17 Comparison of size scale of debris with size of cone cracks.

Other Important Variables: Several important variables have been too difficult or too poorly understood to incorporate into models. These include:

1. *Effect of environment:* There have been many reports that the grinding rate of ceramic materials depends on the pH of the surrounding liquid, on the cation species in the (abrasive) polishing compound, the chain length of hydrocarbons present, the relative humidity of the air in dry grinding, etc. A total picture has not yet emerged, but it appears that one effect of some chemical environments is to decrease K_c values of the ceramic material as much as 50%. Reduction of K_c reduces the energy required to fracture, as may be noted when cutting glass: wet glass fractures much more readily than does dry glass. In grinding processes, the abrasive materials (which are also ceramic materials) would also fracture readily by a reduction in their K_c values, exposing more sharp corners which should increase abrasion rate.
2. *Surface chemistry effects*: Just as oxides form on most metals, so do the surfaces of several ceramic materials react with the environment to form new chemical compounds on the surface. Fischer et al.[15] have shown that for Si_3N_4, sliding in humid air, water, and water mixed with hydrocarbons, a tribochemical reaction with water produces an amorphous SiO_2. This results in a significant decrease of both the friction and the wear rate of Si_3N_4. Further, a reaction between Al_2O_3 and water apparently forms aluminum hydroxides during the

sliding contact. There are many more such reactions, and it appears that some reaction rates are considerably increased by sliding.

3. *Wear particle retention:* No existing wear model for any material accounts for the influence of the retention of loss of wear particles. Wear rates can vary by a factor of 1000 in ceramic materials for the same rate of particle *generation*, depending on the relative amount of reattachment of particles to form the transfer film. In repeat-pass sliding, some or most of the loosened material that might otherwise have been lost as wear debris is crushed into fine particles and recycled through the contact interface. The layer of fine particles (≈1 μm thick, with particles ≈100nm diameter) with large surface area, reattaches rather firmly to the surface, probably by van der Waals and electrostatic attractive forces. Attachment is strongly diminished in the presence of water for example, which may increase the wear rate of ceramic materials by up to six times. Thus the measured wear rate must be taken as the material loosening rate minus the material reattachment rate, and the latter can be a substantial fraction of the former. That there is considerable electronic activity on wearing surfaces may be seen by detecting the emission of ions and electrons from fractured and fracturing surfaces. Cathode luminescence has also been detected (emission of light from an electron-showered surface).

Abrasion, Abrasive Wear, and Polishing[16]

A surface may be scratched, grooved, or dented by a harder particle to produce one or more of several effects. Scratching implies some loss of material, whereas grooving does not.

Scratches and grooves may be no deeper than the thickness of the oxides or other coatings. This may occur if the abrasive particles are softer than the substrate but harder than the oxide (see below), or it may occur if the abrasive particles are very small, e.g., < 1μm (probably not resolved by eye). Groove or scratch widths will probably be of the order of coating thickness (≈10nm). Generally, these fine scratches are not discernible and thus the surface appears polished, that is, the centers of diffraction of the scratches are spaced at a distance much less than the wavelength of light, i.e., <0.1μm. If oxide is progressively removed mostly from the high points of the surface, a surface becomes smoother.

The scratches, grooves, and dents may penetrate into the substrate. Deep scratching will produce debris of the substrate material — metal, polymer, or ceramic. An abrasive particle is abrasive only if it scratches (grooves or dents), and for that purpose the abrasive material must be at least 1 Mohs number harder than the surface in question. (See table of Mohs numbers in Table 2.4 in Chapter 2.)

The hardness differential effect is not abrupt, as can be seen in Figure 8.18. One Mohs number is equivalent to a ratio of 1.3 to 1 in scales of absolute hardness.[16] The rate of material loss by abrasion depends strongly on the shape, orientation, and the manner of constraint of the abrasive.

Shape: Abrasive particles are rarely perfectly sharp. Rather, they have blunted protrusions on them. Three effects may follow, depending on the depth of penetration of an assumed cylindrical protrusion as compared with its radius, as sketched in Figure 8.19.

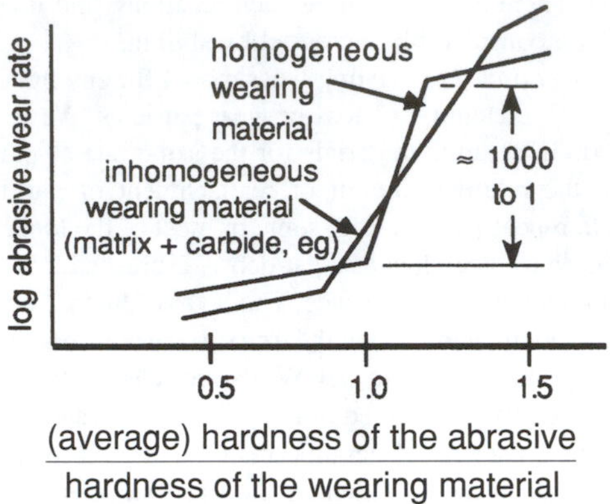

Figure 8.18 Wear rate versus relative hardness of the abrasive.

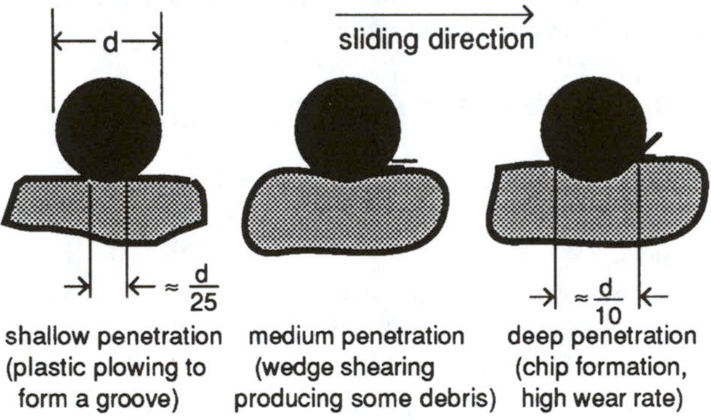

Figure 8.19 The three types of response to depth of penetration (repeated grooving causes fatigue).

A spherical protrusion will behave much the same, except that material may plastically deform to the sides to form ridges, which, if done by repeat passes of abrasive particles in parallel tracks, will also cause wear by low cycle fatigue. This is probably the predominant mode of abrasive wear.

Particle constraint: Fine, anchored abrasives are unusual in practice. Loose abrasives are far more common: they bounce, skid, roll, and cut. The anchored abrasives produce *two-body abrasion,* whereas the action of loose abrasives is called *three-body abrasion.* The fixed abrasives cause about 10 times the wear as the loose abrasives for the same abrasives and the same average pressure in the case of metals, whereas glass and ceramic materials wear faster in three-body abrasion. (Rolling particles are more likely to produce surfaces with diffuse reflection than will sliding, scratching particles.)

Particles in soil are partially constrained by packed surrounding soil, and their abrasive behavior falls between two-body and three-body abrasion. Sandy soils have particles that are about 6 to 6.5 Mohs hard.

In some instances abrasive particles may be crushed between two bodies. The crushed fragments have many more sharp corners than before crushing and are significantly more abrasive. The condition of crushing particles is referred to as high-stress abrasion as distinct from low-stress abrasion. Particle embedment is much more commonly found in high-stress abrasion than in low-stress.

Role of Fluids in Abrasive Wear: Fluids often improve abrasive removal rates over the dry state. Part of the influence of fluids, at least in grinding processes, is to remove heat and debris (swarf). In slow abrasion and in polishing, fluids act primarily to decrease the friction between the abrasive particles and abraded surface, thereby allowing more material removal and less particle embedment for a given amount of expended energy. A second effect of fluid is to lower the fracture toughness of abrasive particles, allowing them to fracture and form sharp edges more readily.

Each type of abrasive material and abrasive operation requires different fluids, as may be found in the directions for use of commercial abrasive systems. For example, oil is better than water for two-body removal of brittle material.

Resistance to Abrasion by Materials: A great amount of work has been done to find abrasion-resisting materials, particularly in the mining and agricultural industries. The primary focus in this work has been on hard materials since hardness is a primary abrasion-resisting property. Generally, martensite is desirable (< 0.6%C martensite has a hardness of 65 Rc or about 800 Vickers Pyramid Number), but for some structural purposes primary martensite is too brittle, and stress-relieving martensite costs money. Steel can be toughened by adding alloys, principally manganese, but this and some other alloys in large amounts retain the softer austenite phase. The formation of carbides in iron and steel alloys resists abrasion because iron carbide, Fe_3C, has hardness of 1200 VPN and the chromium carbides have hardnesses on the order of 1800 VPN. (Iron castings containing significant amounts of carbides are known as white irons, as distinct from gray irons that result from slow cooling, which forms graphite flakes in the matrix.) A concise description of the effects of the myriad of iron-based alloys and microstructures may be found in the book by Zum Gahr.[16]

Laboratory Testing: Abrasion is often erroneously simulated in the laboratory by a larger-scale cutting tool. Some laboratory tests involve sliding the end of metal (specimen) pins on abrasive paper which has a soft backing. Then there are the crusher plate tests, the dry sand–rubber wheel test, the wet-sand rubber wheel test, and many more. The general hope in laboratory testing is to develop an equation or model of wear which would include all of the relevant material properties and abrasive parameters that affect wear rate. Overall, it can be said that wear testers in abrasion more nearly simulate practical wearing situations than do laboratory bench testers in any other segment of the wear field.

(See Problem Set question 8 g.)

Erosion

There are several causes for wear in the *absence* of solid–solid contact. Each cause has a name and these include:

Cavitation[17]: When liquids flow parallel with a flat plate there may be either laminar flow or turbulent flow (or some combination). If liquid is made to flow past a cylinder (for example, a pipe) and perpendicular to the axis of the cylinder just beyond the widest dimension of the cylinder, the momentum of the liquid produces a lower pressure at the solid/liquid interface than in the general vicinity of the system. If the radius of curvature of the cylinder is small and/or if the velocity of the liquid is high, the pressure at the solid/liquid interface may be less than the vapor pressure of the liquid. Bubbles, or cavities, of vapor will form locally and collapse very quickly. The collapse of the bubble may be seen as a flow of liquid with a spherical front toward the solid surface as shown in Figure 8.20.

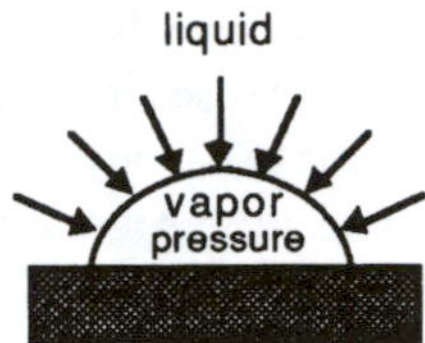

Figure 8.20 Sketch of a collapsing vapor bubble, as in cavitation.

There is sufficient momentum in the liquid to strain the material in the target area. In most regions the strain is much less than the yield strain, but when elastic strains are imposed millions of times in small regions over a surface, local failure of material occurs by (elastic) fatigue. Ship propellers, valves in pipes, and the vibrating cylinder walls of engines are eroded away by this cavitation process.

Abrasive Erosion, Slurry Erosion: When a moving fluid contains abrasive particles, wear will occur. If the velocity is low there may only be removal of oxides, but at high velocity, substrate material is worn away as well. Though there is no clear differentiation between abrasive erosion and slurry erosion the terms often have different uses. Abrasive erosion may refer to low concentrations of solid in liquid, or it may refer to unknown concentrations. The focus is on the liquid phase, and solids are probably considered to be entrained contaminants. By comparison, slurry erosion occurs when a solid–liquid mixture, specifically known as a slurry, causes wear. Generally, such a mixture is called a slurry when the solid phase is the focus of attention and the liquid is simply the carrier. Pumps for moving slurries through pipelines may wear fast: slurries pass through small gaps in the pumps at 100 m/s and more. In some instances, abrasive erosion is desired. Devices are now available, specifically made to propel abrasive particles in water against a hard surface such as concrete and metals in order to cut them.

Erosion by Liquid Impingement: When liquid drops strike a solid surface with sufficient momentum and sufficient frequency, material will be removed from the surface by fatigue. Rain drops erode the polymeric radar domes on aircraft in this manner.

Erosion by (Dry) Solid Particle Impingement[17]: Erosion rate is often measured as mass loss per unit of erodent used. Many erosion variables have been studied. Some are:

1. Solid particles have sufficient momentum to damage solid surfaces. In general, erosion loss rate increases (very) approximately by (particle velocity)^n where n ranges from 2 to 2.5 for metals and 2.5 to 3 for ceramics, and (particle size)^m where m≈3, though it has also been found that n is proportional to particle size, perhaps due to fragmentation. Impingement velocities usually range from 15 to 170 m/s (40 to 150 mph).
2. Impingement of sharp and hard particles at low angles will abrade (cut) soft, ductile material. Material loss by cutting begins very soon after impingement begins. Particles of any shape and hardness, impinging at high angles, will fatigue surface material, causing loss, but the onset of loss is delayed as the material fatigues. Angular particles erode 6061 T6 up to 250 times faster than do round particles, at 70° impingement angle. Hard and sharp particles may embed (up to 90% area) during impingement at high angles.
3. The rules of relative hardness of particle compared to target apply here as in abrasion, except that hardening of the target by cold-work to reduce wear is ineffective, and hardening by heat treatment (which decreases toughness) is mildly effective when eroding particles are hard. Ductility is sometimes more important than hardness in resisting erosion.

   It may be more useful to characterize the hardness of target material in terms of dynamic hardness than static hardness. A 100 μm diameter particle with velocity of 100 m/s has an impact time (for ≈ 5 μm indentation) of 2×10^{-3}s. This produces a strain rate of about 1000 times that in a hardness test which is sufficient to increase hardness by 30%.
4. Figure 8.21 shows how erosion loss varies with impingement angle for four materials.
5. Particle size is a factor (other than in mass and momentum) where target surfaces have different properties than those of the substrate or where corrosion may accompany erosion.

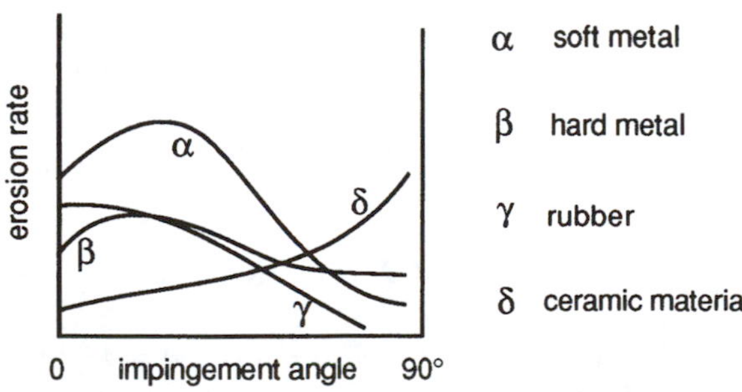

Figure 8.21 Erosion loss rate as a function of impingement angle for four materials.

6. The target may wear away in ripples. Hard particles batter the target, and if it is ductile it may splash outward, causing the formation and loss of platelets. This may occur at sufficiently high strain rate to proceed nearly adiabatically. In two-phase materials the soft phase can be removed first, weakening the support of the harder phase.

Fretting[18]

Tomlinson[19] coined the word "fretting" in 1927. It refers to small amplitude, (apparently) high frequency, oscillatory slip motion between two solid surfaces in contact. The amplitudes range from a fraction of μm to hundreds of μm. Fretting occurs between prosthetic hip joints and bone, gears on shafts, splines, transformer cores, and in many other places. It loosens some joints, seizes up others, and may provide a site for crack initiation.

Small sliding amplitude encourages wear particles to remain in the immediate contact region, which is the characteristic difference between fretting and large amplitude or one-way sliding. The upper limit of amplitude for characteristic fretting has not been found, though the wear rate may increase at amplitudes above 70 μm, and the nature and color of wear debris seem to change at an amplitude above about 100 μm. The actual limit may be connected with the diameter of the microscopic areas of contact.

Fretting wear usually begins at a high rate but levels off after ≈5000 to 10,000 cycles in steel, depending on the ductility of the oxide. Oxide builds around small (asperity) contact areas and carries much of the load. The debris from steel is usually the red, nonmagnetic, hexagonal αFe_2O_3 if the temperature has not exceeded 200°C. At high contact pressure the oxide may appear black because of compaction, and the μ of this form may be low. At higher temperature, the fretting rate decreases.

Very likely, the αFe_2O_3 debris is formed as a lower oxide on the steel and is flaked off, fragmented, and oxidized further. In Al and Ti, bits of metal fatigue off and oxidize, as is seen by metal in the debris. Al and Ti produce black debris.

The influence of frequency on fretting apparently depends on oxygen availability and oxidation rate. Fretted surfaces are often rough, and in some materials this roughness may induce cracks which serve as sites for initiation of fatigue. There appears to be no way to prevent fretting, but reduction of damage can be achieved by reducing slip amplitude or by reducing μ. Surface roughening may help in some cases: it provides escape channels for debris.

PRACTICAL DESIGN

It was implied throughout this chapter, if not stated outright in other places, that wear is so complicated and design tools so sketchy that few mechanical designers can expect to design products for a targeted wear life as readily as they can meet other goals in their products. The following chapters tell more of why this is so and provide aid to engineers with wear problems. The overall message of this book, however, is that wear problems can best be solved by experience and that experience is gained over time by studying worn surfaces and the functioning of wearing machinery, with an interdisciplinary mind.

REFERENCES

1. Bowden F.P. and Tabor, D., *Friction and Lubrication of Solids*, Oxford University Press, Oxford, U.K., Vol. 1, 1954; Vol. 2, 1964.
2. Archard, J. F., *J. Appl. Phys.,* 24, 981, 1953.
3. Archard, J. F., Private communication.
4. Kruschchov, M. M. and Babichev, M.A., *Friction and Wear in Machinery*, 11, 1, 1956. (A translation by ASME from [Russian] *Machinistrinia.*)
5. DeGee, A.W.J. and Zaat, J.H., *Wear*, 5, 257, 1962.
6. Lancaster, J.K., *Proc. Roy. Soc.* (London), 273A, 466, 1963.
7. Welsh, N.C., *Phil. Trans. Roy. Soc.* (London), 1, 257A, 31, 1965.
8. Welsh, N.C., *Phil. Trans. Roy. Soc.* (London), 2, 257A, 51, 1965.
9. Quinn, T.F.J., Trib. Intl., 16, part 1, 257; part 2, 305, 1985.
10. Rhee, S.H. and Ludema, K.C, Leeds-Lyon Conference on "The Wear of Non-metallic Materials," Sept. 1976, Mech. Engr. Publishers, Inst. Mech. Engineers, London, 1976.
11. Muhr, A.H. and Roberts, A.D., *Wear*, 158, 213, 1992.
12. Ajayi, O.O. and Ludema, K.C, *Wear*, 124 237, 1988.
13. Evans, A.G. and Marshall, D.B., in D.A. Rigney (Ed.), *Fundamentals of Friction and Wear of Materials,* ASM, 1981, 439.
14. Lee, K-Y., Fatigue wear mechanism of structural ceramics, PhD thesis, University of Michigan, Ann Arbor, 1993.
15. Fischer, T. and Tomizawa, H., *Wear*, 105, 29, 1985.
16. Zum Gahr, K.H., *Microstructure and Wear of Materials,* Elsevier, Lausanne, 1987.
17. Hammitt, F.J., *Cavitation and Multi-phase Flow Phenomena*, McGraw-Hill, New York, 1980.
18. Waterhouse, R.B., *Fretting Corrosion*, Pergamon Press, New York, 1972.
19. Tomlinson, G., *Proc. Roy. Soc.,* (London), A115 472, 1927.

CHAPTER 9

Lubricated Sliding — Chemical and Physical Effects

Boundary lubrication (chemical function of applied lubricants)
Scuffing (scoring, seizing, galling)
Lubricated wear
Break-in (dynamic changes on sliding surfaces)
— Each of these topics is usually treated separately from the others in scholarly papers. Each is the focus of different academic disciplines, and in each topic the experiments used to verify hypotheses are very different from each other. These differences have tended to obscure the fact that each topic is related to all the others as one continuum of phenomena.

INTRODUCTION

This chapter is about marginal lubrication, here defined by the absence of thick fluid films. Most mechanical items with finite wear life are marginally lubricated, including auto engines. Competition drives consumer products in this direction.

Marginally lubricated surfaces are ever in danger of catastrophic surface failure (CSF) which may end the useful life of sliding surfaces, but their useful lives can be effectively extended by the formation of soft coatings on the sliding surfaces.

There is as yet no way to formally express the adequacy of marginal lubrication in terms of resulting friction, wear rate, or propensity for CSF. Friction, wear, and CSF are often used as measures of adequacy of lubrication, but there is no connection between them. That is, high friction is not always connected with high wear rates, and neither is high wear rate connected with a tendency for CSF.

Several conventional terms require definition:

a. The chemical effects in lubrication are referred to as boundary lubrication and defined more fully in following sections.

b. The catastrophic mode of surface failure is sometimes referred to as scuffing or scoring, or perhaps galling or seizing. Many of these terms are old, poorly understood, and apply principally to ductile metals. Each of these terms has several technical (and subjective) meanings, and each describes different end results. In the interest of reducing the number of terms to be used in this chapter some definitions will be proposed.

   Seizing is a term that describes such severe damage of sliding surfaces that the driving system cannot provide sufficient force to overcome friction: the sliding pairs cease to slide.

   Galling is a process of surface roughening that results from high contact pressure and high traction, at slow speed generally and without any lubricant other than the native oxide and adsorbed gas on the surfaces. Most likely the failure of lubricated surfaces within the first few cycles of contact is similar to the galling process but failure at a later stage is different.

   Scuffing and *scoring* by contrast usually refer to a mode of failure of well-lubricated metal parts. Subjectively they are described as different from galling and seizing, but the initiating mechanism of all may be the same.

   Surfaces that are said to have scuffed have become so rough that they no longer provide their expected function. They may or may not have worn away to any great extent. From this point of view, scuffing is not a wearing mechanism, but primarily a surface roughening mechanism. Scuffed surfaces have a range of characteristic appearances. Some are shiny, some have grooves in them, and some are dull, probably depending on the chemical environment in which they operate.

   Scoring is a parallel phenomenon, sometime manifested as a dull-appearing surface with no obvious roughening. Scored surfaces may only display evidence of overheating of either a lubricant or the metal.

c. *Break-in* refers to the action taken to prepare sliding surfaces for high load-carrying capacity. Generally, new surfaces cannot carry high loads without failure.

Organization of this chapter: The technical literature in the topics of this chapter is scattered among several disciplines and focused rather strongly on metal sliding systems. It is clear that many variables influence the adequacy of marginal lubrication including atomic structure, the mechanical properties of the sliding solids, surface topography, shape of the contact conjunction, sliding duty cycle, and chemistry of the environment, including lubricants. There are many more variables than any discipline can treat and more experimental results than researchers can accommodate and interpret.

The focus on metals is simply historical. The industrial revolution, which by the way, could proceed no faster than developments in lubrication, began before polymers were invented and before temperatures and chemical environments were severe enough to require the high cost of high-quality ceramic materials. More recent research has shown though, that many of the principles of tribology that are applicable to metals also apply to polymers and ceramic materials.

The following sections are short summaries of published thought in each topic, with some critical commentary. The first two sections discuss friction and wear of marginal lubrication. The third is on boundary lubrication; the fourth is on scuffing; the fifth and sixth are on break-in, the latter on the dynamic changes in sliding surfaces, unproven but moderately unified, based on the wide-ranging and specialized research in each topic. The final section offers a few suggestions on designing scuff-resisting sliding surfaces.

FRICTION IN MARGINAL LUBRICATION

Early research showed that there are several separable effects in marginal lubrication. Adequacy of lubrication was measured in terms of friction, and tests were done "by sliding surfaces over one another at extremely low speeds and very high contact pressures so that the incidence of hydrodynamic or elasto-hydrodynamic lubrication is reduced to a minimum."[1] The test specimen shapes of that day were most often the pin-on-disk and the 4-ball geometries.

Sliding of inert liquid/metal pairs produces an increase in friction as temperature increases, shown as curve "a" in Figure 9.1.[2] The rise in friction is attributed to a reduction in viscosity of the lubricant as temperature rises. It is surprising that friction does not rise to very much higher values in such tests. Calculated hydrodynamic films in these tests are of molecular dimensions. These films likely have properties very different from that of bulk lubricant, and they probably do not survive the high temperature achieved on fast-moving surfaces.

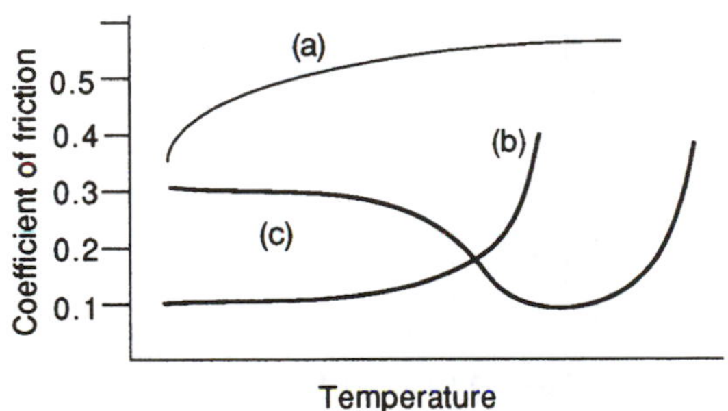

Figure 9.1 Behavior of three types of lubricant over a range of temperature, at very low sliding speed and high contact pressures.

(a) An unreactive lubricant metal pair, in which lubricant viscosity decreases and μ rises as temperature increases.

(b) A fatty acid, or the metal soap of a fatty acid, which melts at a particular temperature and becomes ineffective as a lubricant. For example, stearic acid has MP=65°C and iron stearate has MP=135°C

(c) A liquid lubricant containing a reactive constituent that forms low shear strength compounds on the sliding surface. These reactive constituents are referred to as EP additives, denoting their effectiveness under conditions of "extreme pressure" of contact. Phosphates are effective up to 250°C, chlorides up to 400°C, and sulfides up to 430°C.

The effect shown in curve "b" is simply due to the melting of a solid as temperature increases, as when a grease (for example, fatty acid) becomes a liquid (for example, molten fatty acid). The behavior shown by curve "c" is due to the chemical reaction of a lubricant with a reactive solid, where the reaction rate increases with temperature and the friction decreases. At high temperature the products of chemical reaction often become less effective as a lubricant, so friction rises. The effects shown in curve "c" would not occur when using chemically inert solids as sliding surfaces.

WEAR IN MARGINAL LUBRICATION

The basic mechanisms of scuffing and of steady-state wear may be similar, but phenomenologically they respond differently to changes in overall sliding conditions. Where $\Lambda > 3$ (Λ is the ratio of lubricant thickness to asperity height, to be defined in the section on *The Mechanical Aspects of Scuffing*), there is no asperity contact and yet wear occurs when abrasive materials pass through the system. For values of Λ nearer 1 there will be some asperity contact and some wear, even without abrasives present. Asperity contact can be verified by measuring electrical resistance between conducting sliding pairs. For values of $\Lambda \Rightarrow 0.8$ one can expect severe wear and early failure in nonreactive lubricant/slider systems. This is shown in Figure 9.2 as curve "a."

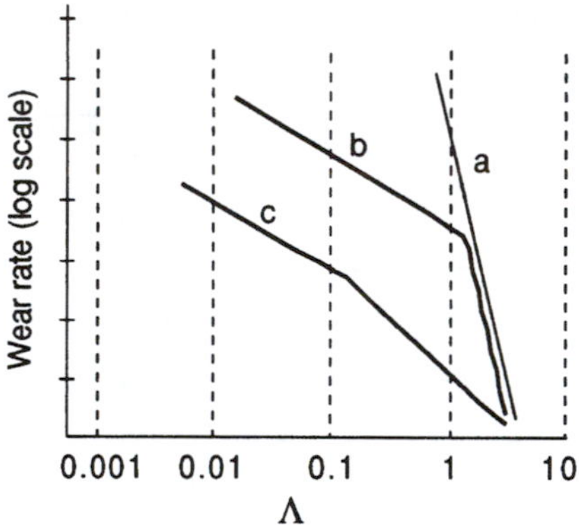

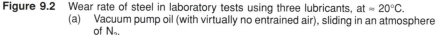

Figure 9.2 Wear rate of steel in laboratory tests using three lubricants, at ≈ 20°C.
(a) Vacuum pump oil (with virtually no entrained air), sliding in an atmosphere of N_2.
(b) Laboratory grade mineral oil (with some entrained air), in air atmosphere.
(c) Commercial gasoline engine oil, in air atmosphere. (Data from Lee, Y.Z. and Ludema, K.C, [ASME] *J. Tribology,* 113, 295, 1991.)

However, few lubricant/slider systems are inert or nonreactive. Modern lubricants contain additives specifically to react with steel (mostly) to form films of low shear strength for higher speed sliding. One poorly understood characteristic of these reactive lubricants is that sliding is required in order to produce effective protective films. Simple immersion and even heating in reactive lubricants have virtually no effect.

The compounds that are formed by chemical reaction in engine oil reduce the coefficient of friction and allow operation at values of Λ well below 1, perhaps to as low as 0.01 for short periods of time. These films include ions of the substrate metal. As these compounds are rubbed off or worn away, some metal ions depart, which constitutes wearing away of metal. (This appears to be one case where the form, at least, of the Archard equation for wear rate is applicable. See Equation 1 in Chapter 8.) The rate of chemical reactivity therefore becomes important. If the desired chemical reaction rate is slow, the rate of generation of protective film may not keep pace with the rate of loss. On the other hand, the chemical reaction rate could be so high that substrate material is corroded away at a greater rate than would ordinarily be lost during sliding. A chemical balance is the goal in the formulation of lubricants for metal cutting, for engines, for gears, etc.

The comparison of the effectiveness of two reactive lubricants with an inert system is also shown schematically in Figure 9.2. (These are the results of laboratory tests described in Figures 9.3 and 9.5.) Among other things, Figure 9.2 shows that the wear rate in boundary lubrication is not some constant fraction of the wear rate in dry sliding, as is often suggested in the literature.

BOUNDARY LUBRICATION

Historical Perspective: Studies of the chemical effects in lubrication began before the year 1900. Early papers are largely anecdotal in nature, describing how test results depended on the source of lubricants, the history of use of the lubricant, the metals in the bearings, and how the bearings were made, to name a few variables. In the 1920s Hardy[4] demonstrated several chemical effects in lubrication and suggested that this mode of lubrication be referred to as "boundary lubrication."

The term boundary lubrication appears to be defined in several ways in the literature. Hardy defined it after observing and measuring the effect of films formed by chemical reaction between a lubricant and a metal. One problem is that boundary films are usually not visible, nor measurable by standard laboratory methods. Thus authors define boundary lubrication in other ways. For example, a machine is sometimes said to be operating in the state of boundary lubrication if its lubricant is known to contain active additives. Others follow the changes of friction as shaft speed decreases or as bearing load is increased: if the coefficient of friction passes to the left through some minimum as shown in Chapter 7, Figure 7.4, then the system must have entered the boundary lubrication regime. Still others will declare that if the coefficient of friction is near 0.1, the system must be operating in the boundary regime. There is some truth in each of these assertions.

Boundary Lubrication in Practical Machinery: There are many papers on boundary lubrication and yet there are few basic principles by which practical lubricants may be selected and bearings designed. Surely the major reason is that boundary lubrication is very complicated, and few authors report research covering a very broad range of experimental variables. Among the great number of published papers, most cover a very small part of the entire field. Many authors work with full-scale machinery but are constrained to study it under either very limited and controlled conditions, or under such general conditions (as in the field) that the data can only be analyzed by statistical methods. Laboratory methods are not entirely satisfying because of the tenuous connection between the behavior of laboratory devices and the behavior of practical machinery. However, some of the laboratory results are worth presenting, if only because laboratory investigations can cover a wider range of variables than is available in practical machinery.

Some Laboratory Results: The experiments discussed here used a cylinder-on-flat geometry with step loading in contrast to the more common ball-on-flat and 4-ball test devices. A sketch of the specimen shapes is shown in Figure 9.3.

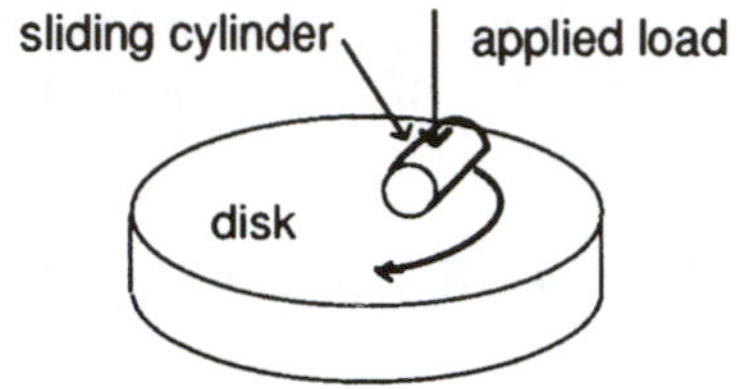

Figure 9.3 Cylinder-on-flat test geometry.

This type of test correlates fairly well with cams/followers in gasoline engines, better than does a ball-on-flat. Figures 9.4 and 9.5 show two ways that tests may be run. Figure 9.4 is a sketch of endurance (time to failure) versus applied load in a lubricated sliding test. Such regular behavior is not often reported but can be achieved by careful design of the experiments.

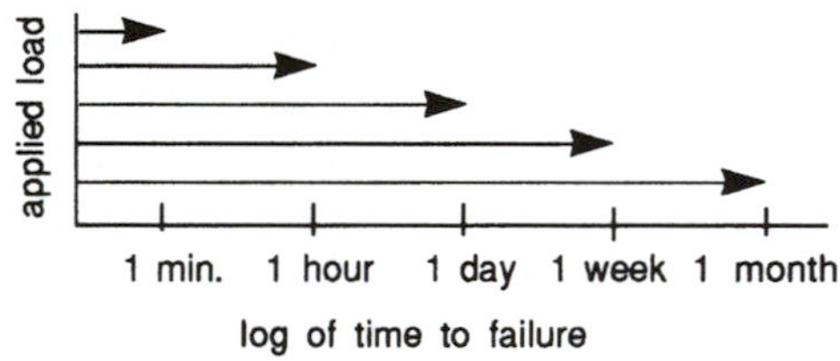

Figure 9.4 Endurance tests.

A considerably shorter test procedure is the step-load test, in which a load is progressively increased after some specified time, as shown in Figure 9.5. The adequacy of lubrication is measured by the load at which surface failure occurs. There is often sufficient correlation between time to failure in the endurance test and the load at failure in the step-load test to encourage the use of step loading.

Step-load tests with steel specimens were used to learn the influence of several variables on lubricant durability. These variables include surface roughness, slid-

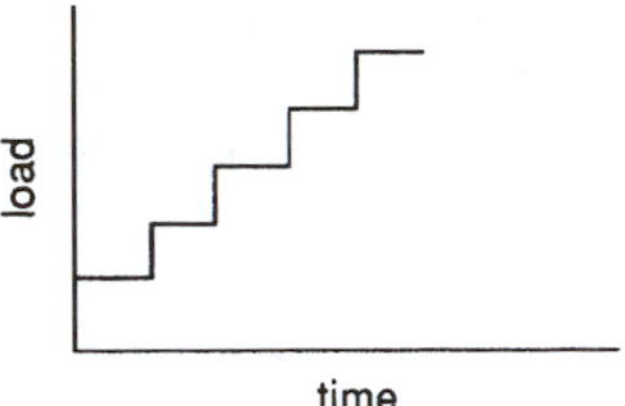

Figure 9.5 The step-load test.

ing speed, metal hardness and microstructure, break-in procedure, and temperature. Further, the chemical composition of the films from these tests was determined, and the mechanical durability of the boundary films was measured.

Composition and Strength of Some Boundary Films: Films form on surfaces of many lubricated mechanical components, but the most studied components have been in automotive engines, transmissions, and other machinery. Most attention has been paid to the widely used oil additive zinc dialkyl dithiophosphate (ZDP). Chemically oriented papers usually describe the films as consisting mostly of oxides and friction polymer. Zinc phosphide, zinc phosphate, zinc sulfate, zinc sulfide, and iron phosphate have been specifically ruled out by one author,[5] but appear to be under discussion by others recently. Most papers show a phenomenological connection between the chemical composition within the lubricant itself and its function as a lubricant. Engine oils contain mostly the original hydrocarbons, paraffinic and/or naphthenic, plus the thermal oxidation products, including alcohols (R–OH), ketones (R–C[=O]–R), acids (R–C[=O]–OH), and esters (R–C[=O]–O–R). (R is a molecule common to each; for example, in methyl alcohol R = CH_3.) In addition to the ZDP additive, there are other additives for acid buffering (to prevent increasing the viscosity of the oil), foam suppression, dirt suspension, etc. The function of an oil as a scuff prevention agent is clearly only a part of the concern of chemists.

The films formed on lubricated sliding steel surfaces include a flake form of Fe_3O_4, upon which is an organo-iron compound (OIC).[6] (For a description of a method to measure film thickness and estimate composition, see the Appendix on Ellipsometry in Chapter 12.) The oxide flakes are a few nanometers thick and less than a micrometer across. The coefficient of friction of a new cylindrical steel slider on a dry steel surface covered with this oxide is about 0.12, whereas for the original dry steel the value is about 0.25.

Relatively little OIC forms when using laboratory grade mineral oil as the lubricant, and it is ketone and acid based. With the addition of ZDP, less oxide forms but the OIC is ester based and forms a thicker layer. It contains up to 15% total P+Zn+S. For the same conditions of boundary lubrication, the plain mineral oil allows between 24 and 80 times more wear (loss of iron) from the cylindrical slider than does formulated engine oil. This is probably because the OIC with P+Zn+S effectively covers the relatively hard Fe_3O_4 flakes.

Figure 9.6 shows that the films require time to develop. The rate of growth is slower for harder disks. Figure 9.6 suggests two distinct layers in the films, but that is the result of the choice of output from the ellipsometer: actually, the lowest part of the film is predominantly oxide, the upper part is predominantly organo-iron, and there is a gradation of composition between.

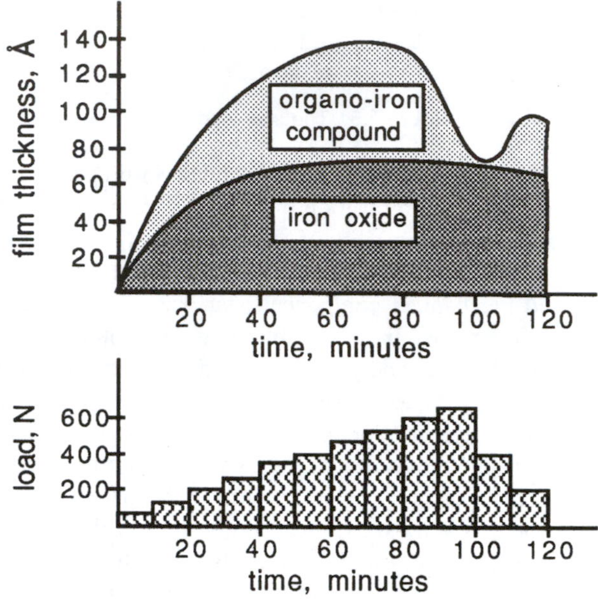

Figure 9.6 Idealized two-film layers on steel disks of 45 R_c, with mineral oil lubricant, with a 6.2mm diameter by 6.2 mm length slider, 0.06m/s sliding speed, step-loading as shown in the bar graphs. (From Cavdar, B. and Ludema, K.C, *Wear,* 148, 305, 1991. With permission.)

The films can sustain limited contact severities, seen as load in Figure 9.6. When the load reaches about 550 N, the films begin to diminish in thickness and the steel surfaces would fail if the load were not relieved. After the load is relieved, the films build up again. This sequence shows that break-in and scuffing are not described by instantaneous values of surface temperature or fluid film thickness, but rather are influenced by the time required to build or wear away the films.

The thickness of oxide film is considerably greater in oil containing 1% ZDP than for either 0.5% or 2% ZDP. ZDP does not increase the rate of film formation, but it does produce a film that will withstand a load about 3 times greater than that sustained by the films formed from mineral oil. Further, the rate of formation of films varies considerably in commercial lubricants over a range of temperature, reaching a maximum at temperatures in the range of 200°C to 300+°C.[7]

The form and type of oxide are important. Monolithic (i.e., furnace grown) Fe_3O_4 is not effective in protecting surfaces. αFe_2O_3, which forms when using either air-saturated oil or water as a lubricant, produces high friction and high wear. ZDP as an oil additive suppresses the formation of Fe_2O_3 but not (flake) Fe_3O_4, at least in laboratory studies.[8]

THE MECHANICAL ASPECTS OF SCUFFING (WITHOUT CHEMICAL CONSIDERATIONS)

Lubricated sliding surfaces sometimes wear away progressively and in a manner that appears to be related to the severity of contact and adequacy of lubrication. Identical systems may on occasion become inoperable rather quickly at almost any point in their expected lifetime. Wear life in the progressive wear mode is often fairly reliably predicted, but the life in the other, the catastrophic mode, is not.

There are several rather detailed conditions proposed for predicting catastrophic failure of surfaces, often referred to as scuff criteria, but none is applicable for design purposes. The consequence is that most lubricated sliding surfaces are considerably over-designed in order to avoid catastrophic failure. This includes a diverse range of mechanical products, such as prosthetic animal joints, heart valves, surgical tools, and machinery of all kinds.

The severely limited scope of existing scuff criteria strongly indicates that our basic understanding of friction and wear is grossly deficient. The principal deficiency is that no criteria incorporates more than one or two of the many relevant variables controlling the events in the sliding interface. This section focuses on evaluation of scuff criteria.

Scuffing is often defined as an adhesive mechanism of wear, implying that the two sliding surfaces become completely welded or bonded together. This cannot be taken as a general definition since sliding surfaces that fail quickly can often be separated without applying force to separate them. The implication may be that the adhesive mechanisms of friction and wear are operative in scuffing as well, but details of the manner by which adhesion occurs are not provided. Further, it is not particularly useful to attribute scuffing to adhesion because in the broader sense, all resistance to sliding and all forms of wearing can be attributed to adhesion. (See Chapter 3 on adhesion.)

Very likely scuffing is not an inevitable sequence from the point of initiation to complete failure. Many sliding systems experience some surface damage early in sliding, that could progress toward failure, but for some reason the damage becomes healed and does not propagate with continued sliding. In other cases the early surface damage does propagate to failure, at various rates.

The three most prominent types of scuff criteria are the Λ criterion, the plasticity index criteria, and the maximum temperature criteria.

The Λ Ratio

The common understanding in the tribology community is that a system is in danger of scuffing whenever the thickness of the fluid film between sliding surfaces becomes less than the average height of asperities on the sliding surface.[9] This condition is expressed in terms of the ratio, Λ, which is defined as:

$$\Lambda = \frac{h}{\sigma} \tag{1}$$

where h is the fluid film thickness as calculated by one of several available equations of elastohydrodynamics (see Chapter 7), and σ is the composite surface roughness of sliding surfaces 1 and 2, defined by

$$\sigma = \sqrt{\sigma_1^2 + \sigma_2^2} \tag{2}$$

There is some validity to the Λ criterion, but with two very significant caveats, namely:

1. The Λ criterion appears to be borne out only in those uncommon systems in which there is very little reactivity between chemically active species in the lubricant and the sliding surfaces.
2. The critical value of Λ is different for every type of surface topography, every type of substrate microstructure, every type of lubricant, and every type of break-in process. Further, these four variables are interdependent.

In summary, the Λ criterion is a useful general indicator of relative lubricating conditions but is not reliable as a design tool for scuff prevention since its critical value may range from about 3 to as low as 0.05.[3]

The Plasticity Index[10]

A scuffing criterion closely related to the Λ criterion is the plasticity index. This concept was developed in several steps over several years, its important point being that scuffing will occur whenever asperities plastically deform to some *small* extent during contact. Precisely why plastic flow should result in scuffing is not stated with much conviction in these papers. Perhaps plastic flow of asperities causes spalling or chipping off of oxide to expose metal to adhesion.

In the plasticity index equations, plastic flow is more likely to occur with materials that are soft but rigid (high E), and where the slopes of the asperities are highest. Intuitively these concepts seem correct, and the theory is partly verified by experiment.

H. Blok began the thinking in 1952.[11] He assumed two surfaces of exactly matching, parallel sinusoidal ridges, of wave length, L, and height, h_{max}, contacting each other on the ridge tops. When these surfaces are pressed together, the ridges are flattened so that full and flat contact between the bodies is achieved. Under certain conditions of ridge geometry and material properties, the pressing together of two surfaces would just produce plastic flow in the ridges. The maximum stress in the material is calculated by:

$$(\sigma_{max})_{crit} = 2(\sigma_{ave})_{crit} = \pi\left(\frac{h_{max}}{L}\right)\frac{E}{(1-\upsilon^2)} \quad \textbf{(3)}$$

The critical value of maximum stress can be taken to be equivalent to the hardness, H, so that:

$$\left(\frac{E}{H(1-\upsilon^2)}\right)\left(\frac{h_{max}}{L}\right) < \frac{1}{\pi} \quad \textbf{(4)}$$

In essence, this equation states that the average slope, h_{max}/L, of the sinusoidal ridges should be less than some value in order to prevent plastic flow in the compressed ridge materials. Tabor calculated the consequences of this condition for tool steel of hardness 800 Bhn and found the critical average slope to be 0.72°. This low asperity slope is rare on practical metal surfaces, suggesting that considerable plastic flow must occur during contact in practical sliding systems.

In 1966 Greenwood and Williamson[12] published a new criterion for the initiation of scuffing (actually, first plastic flow of asperities). They had done considerable and excellent work on the sizes and shapes of asperities in contact and sought to represent asperities more realistically than did Blok or Archard. They assumed two flat surfaces, each of them having spherically shaped asperities on them with average radius, R. The asperities had Gaussian height distribution with a standard deviation *peak* height distribution, σ*, and were sufficiently widely separated from each other so that their strain fields did not overlap. They derived the equation:

$$\Psi = \left(\frac{E}{H(1-\upsilon^2)}\right)\sqrt{\frac{\sigma^*}{R}} \quad \textbf{(5)}$$

where Ψ was referred to as the plasticity index. The quantity, $\sqrt{(\sigma^*/R)}$, can be taken to represent the average slope of asperities, which connects this plasticity index with that of Blok to some extent.

Another view on the plastic deformation of asperities came from Whitehouse and Archard in 1970.[13] They preferred to express the influence of neighboring asperities in terms of an exponential auto correlation function of heights of asperities. Their equation is given in the form:

$$\Psi = 0.6\left(\frac{E}{H(1-\upsilon^2)}\right)\left(\frac{\sigma}{\beta^*}\right) \quad \textbf{(6)}$$

where σ is the standard deviation of height distribution. β* is the correlation distance of the surface topography which characterizes the randomness or uniformity of a surface height profile: a $\beta^* \approx 0$ indicates that surface heights are totally random, whereas a β^* of 1 refers to a flat surface where the surface heights are all interdependent and the same. (Some surface tracing instruments can measure β^*.)

Again, the ratio σ/β^* can be taken as the average slope of asperities. Thus it is seen that in principle, all three of the above equations come to the same apparent conclusion though the slopes are *very* different in magnitude.

Hirst and Hollander did some experiments[14] using the ball on disk configuration to verify the equation of Whitehouse and Archard. For σ they used R_q. They used 18-8 stainless steel, 180 VPN hard, that had been abraded in several ways to achieve a range of values of both R_q and β^*. The slider was a ball of half-inch diameter. They used white mineral oil with 1% stearic acid as the lubricant. A series of tests was done at 95°C and the load was increased progressively until friction increased suddenly. Figure 9.7 shows the approximate ranges of transition loads on $R_q - \beta^*$ axes. The lines are essentially coincident at low values of R_q. Values of constant Ψ are shown on the same axes. Figure 9.7 suggests a connection between specific transition loads and plasticity indices (of Whitehouse and Archard at least) but the connection is fortuitous.

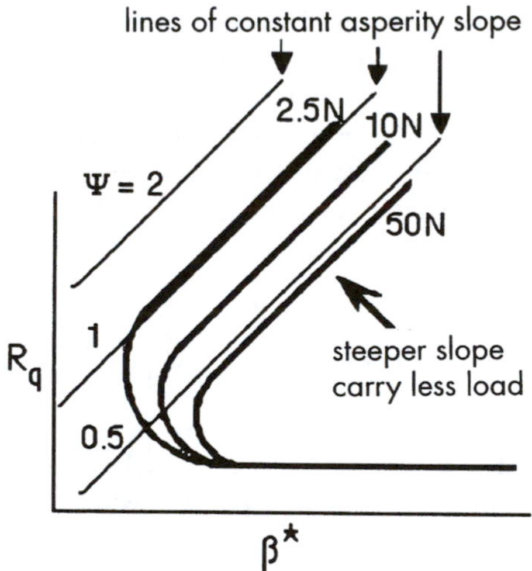

Figure 9.7 Comparison of transition loads and plasticity index values.

Conceptually, it is difficult to understand why specific values of plasticity index have meaning. The transition loads reported by Hirst and Hollander were of the order of 2 N, but conservative calculation shows that only ≈0.2 N will cause significant plastic flow of asperities in their concentrated contact. In fact, in most research a contact stress sufficient to cause global (Hertzian scale) plastic flow is required to cause scuffing. The problem in this topic is that if one assumes that the surfaces are atomically clean then no strain at all is required in asperities to cause scuffing, whereas scuffing resistance of real metals is due to the surface films which are not taken into account in the models.

(See Problem Set questions 9 a and b).

Thermal Criteria

There are several thermal criteria for scuffing, but none is convincing; they are attempts to find the *single* critical surface temperature rise at which some event intervenes to end effective lubrication. These events most often refer to desorption of lubricant or chemical changes in the lubricant.

One thermal theory is based on thermoelastic instability.[15] In this work a local region is thought to heat up by repeated and/or sustained contact. This local region expands and stands higher than the surrounding region. Contact between two such regions on opposing surfaces is suggested to be the site of scuff initiation. Again the connection between high contact pressure and a mechanism of scuffing is missing from the theory.

SCUFFING AND BOUNDARY LUBRICATION

Experimental Work

Considerable work on scuffing has been done under the auspices of the Organisation for Economic Cooperation and Development (Europe).[16] This work produced maps of boundaries between adequately lubricated (partial EHD) and inadequately lubricated sliding of steel on steel, in ball-on-ring tests, over ranges of applied load and sliding speed. A sketch of the form of these maps is given in Figure 9.8.

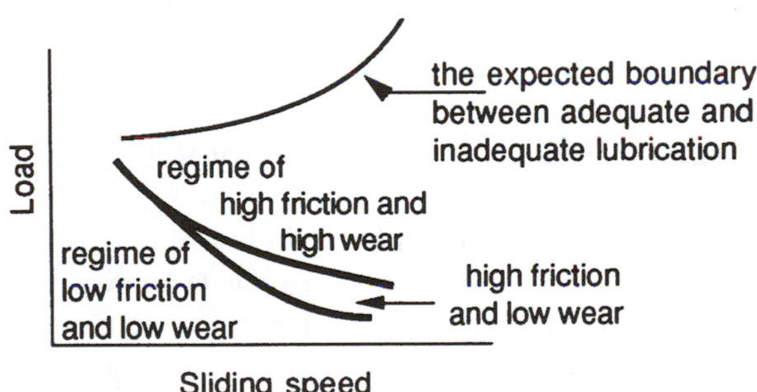

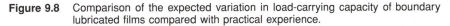

Figure 9.8 Comparison of the expected variation in load-carrying capacity of boundary lubricated films compared with practical experience.

A work following on the method of the OECD showed several additional results.[3] The tests involved wide ranges of lubricating conditions, specimen hardness, and surface roughness, and a few specimen microstructures. The changes in surface roughness, the electrical conductivity, and friction during severe sliding were monitored during the test. Surface roughening was found to change due to two separate causes, namely, plastic flow (probably enhanced by adiabatic heating) and loss of small regions of steel. Friction was low and electrical resistance was high during much of the plastic flow, indicating that direct metal–metal contact or adhesion was not the reason for roughening. The surface roughening occurred very quickly at high sliding speed, coinciding with the low load-carrying capacity shown in Figure 9.8. This seems incongruous since at high sliding speed the fluid film and thus the load-carrying capacity should be higher than at low speed as sketched in Figure 9.8 with a lighter line. The regions above this line are conditions for scuffing and below for safe operation. But this effect may also be due to insufficient time for surfaces to reoxidize after some oxide is removed during sliding.

The loss of small regions or pockets of steel[17] is seen on surfaces in the early stages of scuffing. The size of these pockets is on the order of the grain size of the steel. The base of these pockets showed clear indications of plastic failure, in the form of a lacy-line pattern at 1000× in the scanning electron microscope. Some of these pockets appeared as early as 50 cycles of sliding, suggesting a low-cycle fatigue mechanism of material failure. Further work with ductile metals[18] shows a strong correlation between sliding endurance (contact cycles to the first indication of surface failure) and those tensile properties of metals that correlate with the fatigue life of metals in plastic strain cycling.

Fatigue failure particles on the order of grain size is also seen in ceramics.[19] This type of failure is very sensitive to the ratio between traction stress and normal stresses (i.e., the coefficient of friction) on the surfaces (seemingly a characteristic of brittle materials). Fatiguing is a prominent mode of surface failure and must be included in future criteria for catastrophic failure of surfaces.

One important conclusion that can be drawn from the above quoted work is that the several scuffing criteria mentioned are all *condition criteria*. That is, when certain (static) conditions are met, scuffing will occur. Surely in the case of lubrication with chemically active liquids and where fatiguing is a prominent mode of substrate failure, there is also a significant, if not a primary, *history-dependent* component.

Further Mechanical Effects of the Boundary Lubricant Layer

Surface roughness has an effect on break-in in addition to that expressed in the value of Λ. With soft steel it was found in some laboratory experiments that, under the conditions of the test, proper breaking-in by sliding requires a specific initial surface roughness of about 0.1 μm R_a. Smoother and rougher surfaces failed quickly as shown in Figure 9.9.[8] It appears that the optimum surface roughness is one in which the asperities plastically deform at a rate that is too slow for fast progression to low cycle fatigue failure of surface metal, but at a rate sufficient to accelerate the formation of oxides. The surface roughness of the original smooth and rough surfaces stayed the same throughout most of the tests. The surfaces with intermediate surface roughness became smoother.

Dry Boundary Lubrication

Boundary lubricants are usually thought to be applied to or inserted between sliding pairs, producing nonsolid boundary films. Actually, useful and protective films are often formed on solid surfaces simply by reaction with gases and vapors: they prevent high friction and wear. Many metals are coated with oxide films,

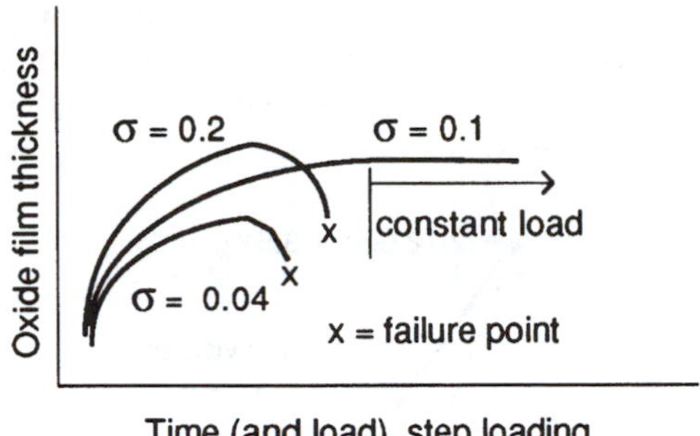

Figure 9.9 The film thickness change during a step-load test, as influenced by the surface roughness, R_a. (Adapted from Kang, S.C. and Ludema, K.C, *Wear,* 108, 375, 1986.)

which form in air, some of which are protective. Rhenium oxide,[20] some iron oxides, and copper oxides on some copper alloys are examples. Other oxides are not particularly protective, such as the oxides of chromium, aluminum, and nickel. Another protective boundary film is formed from gaseous hydrocarbons, which by catalytic reaction of a metal surface will deposit graphite upon that surface.[21]

(See Problem Set question 9 c.)

SURFACE PROTECTION WHEN $\Lambda < 1$ — BREAK-IN

The surfaces described in curves “b” and “c” in Figure 9.2 survived but wore slowly though Λ was small. The reason was that the step-load sequence allows a conjunction to begin at high values of Λ allowing a protective film to form before a high load is applied.

Some practical machinery and devices will survive starting from new with full design load. This is most often the case with low-cost or over-designed items. Makers of large and more expensive mechanical components often break in machines by operating them gently at first, or in some cases with a special oil, sometimes containing a fine abrasive. Each strategy has its own purpose. An abrasive compound in oil simply laps the sliding surfaces into conformity. Some break-in oils contain a more chemically active additive than normal to accelerate the formation of films in regions of high contact stress. However, if these oils are left in the system they might cause excessive corrosion.

The effectiveness of break-in may be seen in some laboratory tests. Tests were done using the geometry of Figure 9.3 to determine how long a lubricated surface would survive when loads were applied by two methods, namely by progressive loading to some final load, and by immediate application of the same final load, both at constant speed.[22] The results are shown in Figure 9.10. At a contact pressure of 0.2 GPa the survival time with progressive loading is 20 times that with immediate loading. (Progressive loading to the target value requires less than 5% of the total time.) At 0.8 GPa there is a tenfold difference, and at 1.5 GPa the difference is only four times. Results of both tests converge to a single point at a contact pressure in excess of 2.1 GPa or ≈ 3.5Y, where Y is the tensile yield strength of the metal. The benefits of a break-in procedure appear to diminish as loads increase. There is also a contact pressure below which the value of Λ is great enough to totally avoid wear.

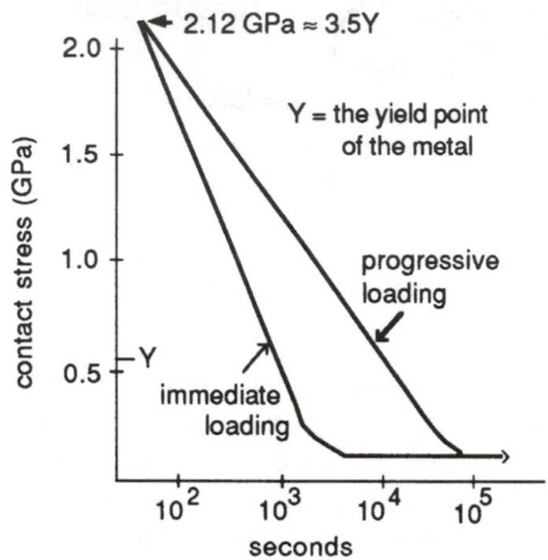

Figure 9.10 The durability of steel sliding surfaces, with lubricants containing reactive constituents, comparing systems where full loads were applied immediately and progressively. (Adapted from Lee, Y.Z. and Ludema, K.C, *Wear,* 138, 13, 1990.)

DYNAMICS OF BREAK-IN

General Conditions

In general, new surfaces are at risk until some protection develops. Most surfaces are not deliberately coated before start-up, but to prevent failure compounds are sometimes applied to new surfaces, which are referred to as break-in coatings. These are often phosphates of iron, manganese, or zinc. Their exact role has never been determined. Some authors suggest that these compounds have rough surfaces which trap and hold lubricant until other protective films can develop. Others suggest that these compounds simply have lower shear strength than does a substrate metal, which then function as solid lubricants. It is also likely that pre-applied break-in coatings prevent or retard the formation of boundary films from chemical sources in the lubricant.

The dynamics of break-in have not been well studied. Break-in is often assumed to be a surface smoothing effect, which would be expected to increase Λ. The problem with that view is that some surfaces become rougher during break-in than the original, and they function very well also after break-in. Perhaps some balance of events occurs as illustrated in Figure 9.11.

In it are plotted two competing sequences for two cases, namely, the protective capability of chemically formed films versus what the sliding surfaces require to survive. Both change with time of sliding. A surface as manufactured has some

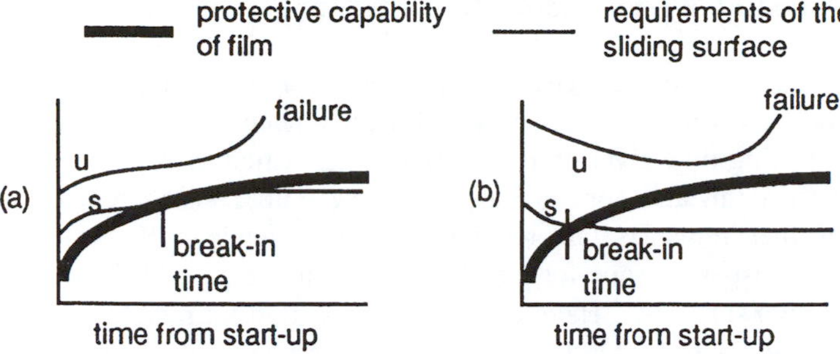

Figure 9.11 Showing the comparison of protective capability of films (which increases with time) versus the requirements of the sliding surfaces, each graph showing surfaces that can (s) and cannot (u) become satisfactorily protected by the developing surface film. (a) Surfaces that become "worse" with sliding. (b) Surfaces that become "better" with sliding, for a time.

initial but unquantified requirement in lubrication, according to its surface condition, hardness, and, doubtless, other properties. Following the relatively unprotected start-up, sliding surfaces change in some way (for example, roughness), which usually increases the requirements for lubrication, as shown in Figure 9.11a. If the protective capability of films overtakes the changing surface requirements, the surfaces will survive. Otherwise the surface will fail. Figure 9.11b shows a case where surfaces improve for a time, but would not continue to do so in the absence of a protective film.

Competing Mechanical and Chemical Mechanisms

Another view of the competition between tendency to scuff and development of protection by a boundary film may be seen in Figure 9.12. The condition of boundary lubrication is one of low Λ, with some, if not a significant part, of the load carried by high stress contact of randomly distributed asperities. This is shown in Figure 9.12a. As sliding continues, debris may be generated or the surface may change in such a way that some points of high stress contact gather or agglomerate as shown in Figure 9.12b. Finally, when damage (streaking) from one asperity site extends into the region of the next following high contact stress site, damage propagates beyond control and scuffing occurs. The third stage is shown in Figure 9.12c.

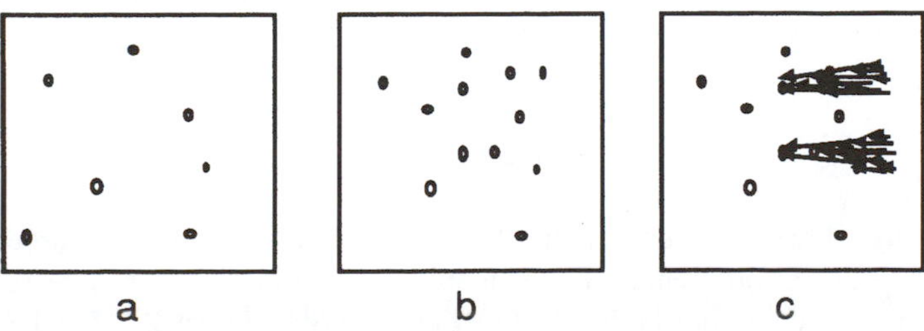

Figure 9.12 Possible progression, from a to c of surface change until scuffing occurs.

Joint Mechanical and Chemical Interaction

In the absence of serviceable scuff criteria a scuff map could show the interaction between some variables and scuff tendency. Figure 9.13 could be called a scuffing map given in the terms of the plasticity index, namely, R_q and β^* in that it delineates zones of high and low scuff resistance, showing other conditions that influence scuff resistance. Scuffing tendency is clearly found to be related to asperity slope for metals that are lubricated with inert liquids, but only for rough surfaces.[10] There is no such connection for metals lubricated with reactive liquids. Curiously, there appears to be no connection between scuffing tendency and slope for the smoother surfaces as shown in Figure 9.10. This observation is parallel to the general impression that there exists some optimum surface roughness, above and below which scuffing is more likely than at the optimum roughness. For decades this was thought to arise from oil being stored in cavities for use under severe conditions, but this concept does not explain the poor functioning of very rough surfaces, which store much oil!

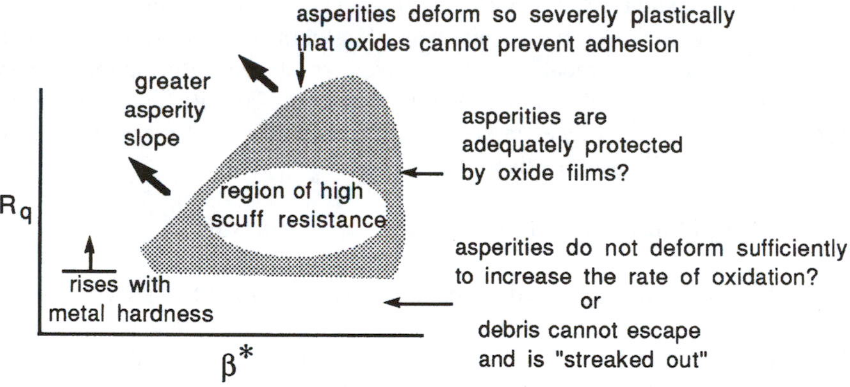

Figure 9.13 A scuff resistance map.

The high scuff tendency for very smooth surfaces has not been adequately explained. There could be two reasons: either the asperities on smooth surfaces do not plastically deform sufficiently to accelerate the growth rate of beneficial oxides, or wear debris is trapped and is severely ironed. Perhaps both occur simultaneously.

(See Problem Set question 9 d.)

PERSPECTIVE

The literature on the lubrication of inanimate machinery is very large but the products of our efforts remain rather feeble. By contrast, God has designed the 100+ bearings within our bodies to function beautifully over $2 - 5 \times 10^6$ cycles without specific attention. Actually they are supplied from those materials that we so readily consume and refer to as nourishing.

The bearing surfaces of our joints are nonconforming and composed of articular cartilage with a roughness of between 2 and 20 μm Ra. This cartilage is porous, filled with between 70% and 80% synovial fluid, which is hyaluronic acid (straight polymeric chains) in a base of dialysate of blood plasma. In vertical loading the effective Young's Modulus of the cartilage is about 10 KPa, which decreases to as low as 0.1 KPa over time, not primarily in a visco-elastic manner but by "weeping" fluids from the pores: this acts to "damp" straining excursions.

In sliding, values of μ in the range from 0.002 to 0.05 have been measured and are due to a combination of two mechanisms:

1. At low sliding speeds, a chemical mechanism from the acid provides boundary lubrication
2. At higher sliding speeds, weeping occurs as contact pressure advances across the surface of the cartilage, providing the fluid for hydrodynamic lubrication.

PROGNOSIS

The topics of this chapter constitute much of the unfinished work in tribology. All of the sciences and arts of tribology are required to design lubricated systems that avoid catastrophic failure, and do so economically.

A much larger literature could have been cited in support of various points in this chapter since very many highly competent researchers have worked in and near these topics. The unfinished nature of these topics is seen only because of the high quality of previous work.

Though there are no applicable scuff criteria available, it is possible to set down a few guidelines for designing for scuff resistance. These guidelines begin with the recognition that the scuff resistance of surfaces changes from the time they are first put into service and continues to change as the duty cycle of surfaces changes during surface lifetime.

One very important component of lubricated sliding systems is the wear particles that form during sliding and/or are inserted from outside. Some of the internal particles are the oxides and other chemisorbed compounds on surfaces, and some develop from substrate material. The brittleness and stickiness of these particles are important properties. For example, brittle particles are likely to be smaller than ductile particles and are more likely to be pulverized than to grow than are ductile particles. Brittle particles are also less likely to be sticky than ductile particles, but a given volume of brittle particles is likely to damage a surface more than will ductile particles. But again, brittleness and ductility are relative terms: most debris is probably made up of a mixture of brittle and ductile particles.

A consideration in designing for scuff resistance is the mechanical constraint of the sliding surfaces. If two surfaces can separate to accommodate momentary growth of debris particles, the particles are likely to do less damage than if the surfaces cannot separate. An example of the latter would be cylindrical plungers or a shaft in a snug-fitting cylinder.

Other considerations include:

1. Whether two sliding surfaces, in repeat-pass sliding, will follow the exact same path each time they pass
2. Whether sliding is always in the same direction or reciprocating
3. Whether the contact conjunction *shape* remains the same at all times
4. Whether the sliding surfaces operate within a severely vibrating environment.

Very likely few of these conditions can be reasonably well modeled, requiring a significant measure of experience with specific products to design successfully for scuff resistance and successful break-in.

REFERENCES

1. Bowden, F. P. and Tabor, D., *Friction*, Anchor/Doubleday, 1973, 120.
2. Bowden, F. P. and Tabor, D., *The Friction and Lubrication of Solids,* Cambridge University Press, Cambridge, U.K., 1950.
3. Lee, Y.Z. and Ludema, K.C, (ASME) *J. Tribology,* 113, 295, 1991.
4. Hardy, W. B., *Collected Works*, Cambridge University Press, Cambridge, U.K., 1936.
5. Beerbower, A., Boundary lubrication, *ASLE Trans*., 14 90, 1971.
6. Cavdar, B. and Ludema, K.C, Part I: *Wear*, 148, 305, 1991. Part II: *Wear*, 148, 329, 1991. Part III: *Wear*, 148, 347, 1991.
7. Choa, S.H., Ludema, K.C, Potter, G.E., DeKoven, B.M., Morgan, T.A., and Kar, K.K., A model of the dynamics of boundary film formation, *Wear*, 177, 33, 1994.
8. Kang, S. C. and Ludema, K.C, *Wear*, 108, 375, 1986,
9. Tallian, T.E., McCool, J.I., and Sibley, L.B., *I. Mech. E*, 180, 3B, 238, 1965.
10. Park, K.B. and Ludema, K.C, *Wear*, 175, 123, 1994.
11. Blok, H., Proc. Gen. Disc. on Lubrication, *I. Mech. E.*, p. 14, 1937.
12. Greenwood, J.A. and Williamson, J.B.P., *Proc. Roy. Soc*. (London), A295, 300, 1966.
13. Whitehouse, D.J. and Archard, J.F., *Proc. Roy. Soc*. (London), A316, 97, 1970.
14. Hirst, W. and Hollander, A.E., *Proc. Roy. Soc*. (London), A337, 379, 1974.
15. Johnson, R.R., Dow, T.A., and Zhang, Y.Y., *ASME, J. Tribology*, 110, 80, 1988.
16. DeGee, A.W.J., Begelinger, A., and Salomon, G., *Proc. 11th Leeds-Lyon Symp. on Tribology*, 105, 1984.
17. Xue, Q.J. and Ludema, K.C, *Proc. Conf. on Wear of Mat'ls*, ASME, 499, 1983.
18. Kim, K.T. and Ludema, K.C, Low cycle fatigue as an initiating mechanism of scuffing, *ASME. J. Tribology,* 117, 231, 1995.
19. Lee, Y.K., PhD. thesis, University of Michigan, Mech. Engr. Dept., 1993.
20. Peterson, M.B. and Florek, J.J., *Sliding Characteristics of Metals at High Temperature*, University of Michigan, Engineering College Industry Program, publ. IP357, May 1959.
21. Lauer, J. and DuPlessis, L., Relation between deposition parameters, structure and Raman spectra of carbon overcoats on simulated magnetic storage disks, (STLE) *Trib. Trans,* 49, 346, 1993.
22. Lee, Y.Z. and Ludema, K.C, *Wear*, 138, 13, 1990.

CHAPTER 10

Equations for Friction and Wear

IN ALL TECHNICAL FIELDS THE HIGHEST ACCOMPLISHMENT IS TO FORMALIZE KNOWLEDGE INTO MATHEMATICAL (EQUATION) FORMAT. THIS ACTION SERVES TWO PURPOSES, NAMELY, TO PROVIDE EQUATIONS WHICH ENGINEERS CAN USE IN PRODUCT DESIGN, AND TO ADD PURPOSE AND DISCIPLINE TO RESEARCH. EQUATIONS FOR FRICTION, FOR SCUFF RESISTANCE, AND FOR WEAR LIFE ARE IN A MOST PRIMITIVE FORM AT THIS TIME.

INTRODUCTION

Designers need equations for all phenomena that control the cost, function, and reliability of the products they design. There are useful equations for fluid flow through pipes, for the energy required to achieve mechanical action, and for the voltage drop across a resistor. But there are no broad and "user friendly" equations for predicting the frictional behavior (vibration potential for example), the adequacy of lubrication under severe contact conditions, or the wear life of mechanical products. These quantities are therefore estimated, postponed, or avoided in the design process.

Indeed, there are several equations in the general area of tribology, for contact stresses (Chapter 5), temperature rise on sliding surfaces (Chapter 6), and hydrodynamics (Chapter 7). These equations have been immensely helpful to our developing technology, particularly in designing high-load and high-speed bearings of all types. However, none of these equations, nor the methods used to obtain them, has been successfully extended to the topics of friction, scuffing, or wear rates.

The rate of progress in the development of useful equations is so slow that most readers of this chapter will likely not benefit from them during their careers. The justification for this prediction follows, but the major point of this statement is to encourage designers to adopt alternate methods in design.

WHAT IS AVAILABLE

The two equations for friction in Chapter 6 have been shown to be inadequate, Equation 1 partly because it does not consider the plastic flow of asperities and Equation 7 because of our inability to characterize the shear strength of adsorbed films on surfaces. The equations for wear in Chapter 8 have been shown to be inadequate as well, either because they include undefinable variables or are based on erroneous material-removal concepts. Equation 1 of Chapter 8 requires special attention because of its widespread use. It is attributable to J.F. Archard and is reproduced here: the time rate of wear, Ψ, due to adhesion, is given as:

$$\Psi = k\left(\frac{WV}{H}\right) = \left(\frac{N \times \frac{m}{s}}{Pa = \frac{N}{m^2}}\right) = \frac{m^3}{s} \qquad \textbf{(1)}$$

where W is the applied load, H is the hardness of the sliding materials, V is the sliding speed, and k is a constant referred to as the wear coefficient.

This equation was published in 1953 and is based on the methods of solid mechanics alone. The one quantity related to materials in this equation is hardness, which Archard knew to be an inadequate and misplaced parameter because of the results of experiments he was engaged in at the time. However, it is an equation that many people have confidence in, a few because it is applicable to their particular product, some because it is widely quoted, and some because the units make sense.

Equation 1 is inadequate because it incorporates only three of the likely 30+ parameters needed for completeness (a point to be made later). The imprecision of this equation may be seen in the very large range of values of k, extending from 10^{-4} to 10^{-9}. No one is able to predict k for any particular application to better than one order of 10 accuracy. Designers need predictions in the range of accuracy of ±10%. However, most defenders of the equation apparently assume that when other parameters are properly identified, these can readily be placed within the framework of Equation 1, a dubious hope. Furthermore, it is asserted, most of the equations that engineers use are only approximations anyway, which are useful until something better is available. The point of acceptable approximations is, of course, arguable.

There is no way at this time to predict the final forms of useful equations for friction and wear, but there are distinctions between origins and types of equations that are worth discussing. The next section discusses these points using terminology that is somewhat confused in engineering. The identification of the parts of Equation. 1, to be used in the following paragraphs, is shown in the sketch:

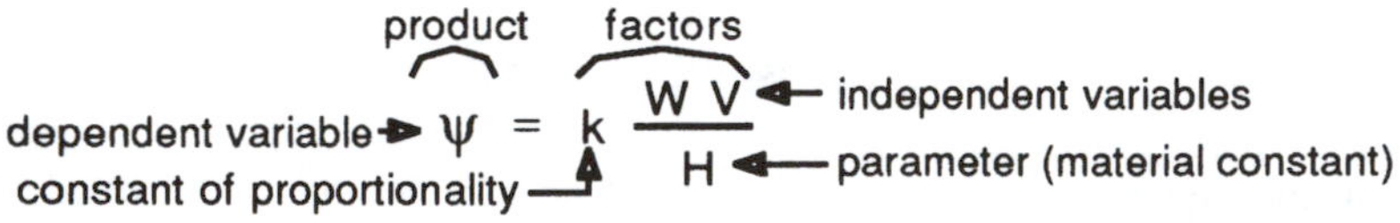

Some equations also contain factors relating to physical dimensions of system parts which will be called parameters as well.

TYPES OF EQUATIONS

Three major types of equations will be discussed: fundamental, empirical, and semiempirical. For completeness a type of equation known as a model will also be discussed.

Fundamental Equations

The most useful type of equation is derived strictly from a knowledge of the controlling variables. For example, the deflection of the end of a cantilever beam, $\delta = PL^3/3EI$ places all of the important and relevant variables into one equation for engineers to use. (P=the applied load, L is the length of the beam, E is the Young's Modulus of the material in the beam, and I is the section modulus of the beam cross section.) With time and usage the engineering community gained confidence that this equation describes the behavior of a beam within a margin of error of ±5%. All variables are readily measured, none is omitted, and each is independent of the others.

Before equations of this type were developed engineers kept records of the deflection of various beams with various applied loads and when a new situation arose, the engineer would extrapolate from or interpolate between known conditions to predict beam behavior in the new condition. Perhaps some empirically constructed equations could have sufficed had not fundamental equations appeared, but they would never have had the predictive capability of fundamental equations.

Fundamental equations require or contain no adjustable constants that represent some heretofore unknown phenomena. However, neither should they be derived from extensive lists of likely variables. Such exercises appear now and then in the literature. They begin with a long list of variables, each represented by a different symbol. Wear rate then is the product of all of the named variables. To assure that the importance of each variable is properly represented, each is raised to a separate exponent so that the final form of the equation is, for 26 variables:

$$\Psi = A^a B^b C^c D^d E^e F^f G^g \ldots . Z^z$$

Further in the general case each of the variables and exponents is independent of all others. Some 10 million experiments are required to find numerical values for the exponents alone.

Empirical Equations

Empirical equations are those developed from experimental data by fitting curves to the plotted data or by estimating equations by least squares or other methods. For some phenomena, empirical equations are necessary because some factors for inclusion in fundamental equations have not yet been determined. For example, the life, T, of a cutting tool in a lathe has been found by experiment to depend on the cutting speed, V, in a logarithmic manner. This relationship was not previously known. The best relationship between these variables is $VT^n=C$ where n and C are taken from experiments. Equations of this type would suggest a relationship that is valid over an infinite range of cutting speed, but recall that an empirical equation is only an approximation of fact. Actually, the equation is valid only over the range of tests, which, for tool life, is usually selected to cover some part of the practical range of cutting conditions.

Any one set of constants, n and C, applies to very specific conditions of depth of cut, feed rate, tool shape, tool material, material being cut, type of coolant, and perhaps machine-related variables such as vibration characteristics.

To generalize the equation these other variables should be included. The common method of doing so is to conduct tool life tests with every variable fixed except the one of interest. The slope of the curve of tool life versus depth of cut, d, for a particular cutting speed is then obtained. The same type of data is obtained for the influence of tool life versus feed rate, f, etc. The slopes, a and b, then become exponents in an equation of the type $VT^n f^a d^b = C_1$, provided there is a logarithmic relationship between both feed rate and depth of cut with tool life. These exponents are usually not precise because the data used to obtain them are derived from tests using very small ranges of the other variables.

To generalize still more there should also be some expression to represent tool shape and coolant, but these have been found to be rather tedious to work with. By general agreement in the community of users of cutting tools, one ceases to make the equation more complicated beyond some point, supplementing the short equations with tables of m "constants" for the remaining variables.

It should be noted that the equation $VT^n f^a d^b=C_1$ is actually not a wear rate equation. Rather it expresses a cutting time after which the tool is useless, i.e., its cutting edge has rounded off and no longer cuts. The equation should be called a material-removal-time equation, or a performance model, and some wear equations in the literature are of the same type. The actual wear (material loss) rate of tool material is nonlinear in time, increasing toward the end of the test. Even then, a simple expression of persistence of the tool in cutting is inadequate in engineering practice. Often the more relevant condition for stopping a cut is the deteriorating condition of the surface of the part being cut: surface roughness and residual stresses usually increase somewhere in the last half of tool life, indicating that tool life should be evaluated according to its useful cutting life rather than the time over which it continues to remove material. The development of a good tool life equation from the fundamentals has not yet been achieved despite 70+ years of effort.

(See Problem Set questions 10 a, b, and c.)

Semiempirical Equations

Equation 1 is an example of a semiempirical equation. It might also be referred to as a semifundamental equation. In essence, with this type of equation an author sets down some variables and parameters that should or are considered to govern wear rates. However, some likely important parameters and variables are left out for some reason. Now, to arrive at a credible product of all known factors an experiment is conducted and a factor known as a constant of proportionality is determined. A budding fundamental equation has thereby been aided by some empiricism.

Models

There are three prominent forms of models: word, pictorial, and mathematical. Word models are descriptions of phenomena or behavior of materials, etc. Pictorial models are sketches (etc.) and/or a series of sketches describing the functioning of some device or phenomenon.

Mathematical models are equations that simulate, or describe the response of, some entity of unknown internal composition ("black box") to some input variable. If the black box is a complicated mechanical system, consisting of springs, masses, and dampers, both an input time-varying force and the output frequencies and amplitudes are measured and the nature of the coupling, or transfer function, is written in mathematical form and is called a model of the system. The same is done with electronic circuits consisting of resistors, capacitors, inductors, diodes, et al., and with chemical reactions.

A model can never be a complete description of a system, but methods have been developed to improve their utility. One problem that arises in measuring the input and output variables of a system is that the measuring system itself introduces error into the results. Systems are identified by testing under conditions that will separate the errors of the system from the errors of the measuring system. Further by a statistical wave-form analysis, some random behavior events can be separated from real system behavior.

TOWARD MORE COMPLETE EQUATIONS FOR FRICTION AND WEAR

The claim that there are no useful equations for friction and wear for designers may appear to be exaggerated given that very many equations can be found in the literature. Apparently then, one needs only to select the correct one for each problem at hand. This is a futile expectation since it is highly unlikely that the problem at hand was the subject of the research from which the candidate equations emerged. Results from wear tests can rarely be extrapolated to other situations.

Though specific equations in the literature are rarely applicable there might be some truth in each of them. Perhaps a composite equation could be assembled, or perhaps the preponderance of use of some parameters might indicate the importance of one material property over another. It was for these purposes that a search was conducted for a common theme in wear equations.

The Search

A search was done in the 4706 papers in *Wear Journal* from 1957 to 1990, and 751 papers in the proceedings of the conferences on Wear of Materials from 1977 to 1991.[1] Many more journals could have been scanned but these two sources are specifically devoted to wear, and they are well reviewed. Papers from other journals were also analyzed if they were referenced in one of the primary papers.

The great majority of papers were discussions and logical descriptions of how wear progresses, which could be called word models. Most of these are accompanied by micrographs, electron/x-ray spectra, and other evidence of wear damage.

An intermediate form of tribological information is embodied in wear maps. One type of wear map has been published by Lim, Ashby, and Brunton[2] and is shown in Figure 10.1. These authors assumed that wear is influenced primarily by stress and temperature rise during sliding, which is sometimes true. They divide the practical domain of the reduced-stress versus reduced-sliding-speed plot into regimes in which particular mechanisms are thought to prevail. This approach is still under development and should not be dismissed. However, it suffers from the same limited perspective as does Equation 1. Many transitions in wear are only mildly sensitive to stress intensity and/or temperature, but more sensitive to such omitted variables as surface chemistry and number of stress cycles.

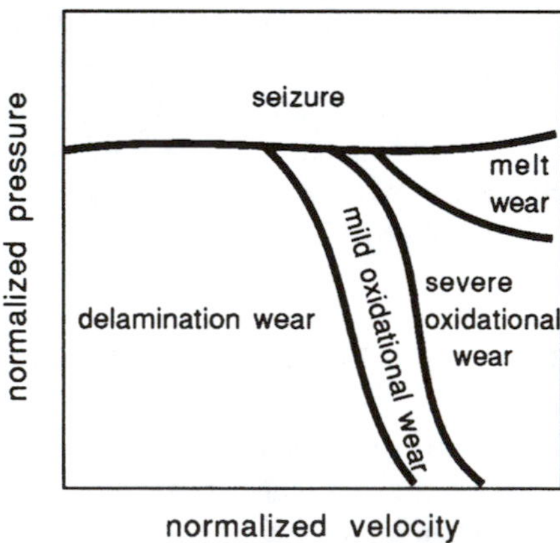

Figure 10.1 A map of wear mechanisms related to conditions of sliding.

Over 300 equations were found for friction and wear. The exact number is difficult to state because some equations are very small revisions of previous equations. The equations were scanned to determine whether they converged upon a few most desirable variables. In principle, it would seem that one (or a few) good equation(s) could be condensed from the great number of published equations, provided the true importance of a few central variables could be determined. One of the first discoveries was that many of the equations appeared to contradict each other and very few equations incorporated the same array of variables. It is common to find, for example, Young's Modulus in the numerator of some equations and in the denominator of others. It therefore seemed obvious that a simple tabulation of the uses of each variable would not indicate the importance of that variable in the wearing process. A method was necessary to separate out those equations that *properly* represent variables from those that do not. Perhaps from these equations a reasonably authoritative hierarchy of factors could be established to help develop fundamental equations.

Analysis of Equations

The equations were evaluated against four criteria for the purpose of finding the most useful of them. The following criteria were applied:

a. Historical significance
b. Applicability
c. Logical structure
d. Nature of supporting information, especially from experiments

Historical significance: The equations of authors who published a progression of thought in the same general topic were given higher credence than those of authors who published only once. For example, of the 5137 authors named on the 5467 papers, 3257 were named on papers only once, 810 were named on two papers, and 362 on three papers. Fewer than 100 authors published more than 10 papers, and only 291 have remained in the field more than 5 years.

A second criterion in this category was to see how peers regarded a published work by analysis of reference lists. This procedure favors older papers, which was not intended, but it adds the opinions of a broad range of authors. (It is particularly instructive to see how an author responds to commentary on his work, and to note how often an author quotes his own work beside the work of others.)

It was also noted that papers older than 15 years old are seldom referenced, except sometimes in a group of references taken from another, fairly recent paper. Also, mathematical papers are referenced many more times than are those containing micrographs and other information tedious to comprehend. An analysis of significance based on these criteria required some judgment.

Applicability: Equations contain many different variables. Most often the familiar variables are used, such as hardness, Young's Modulus, etc. However, many equations contain variables that are not readily definable, or are only available from experiment. Examples are grain boundary strength or atomic damping factor or surface stiffness. Rarely do the authors of such variables follow up and measure these quantities themselves, and neither does anyone else. Such equations have limited use.

Many equations have limited use also because they only show relationships between variables without providing information on the resulting friction or wear rate.

Logical structure: Some equations are built on strings of poorly rationalized assumptions. It is difficult to show precisely where, in such equations, the overall argument departs irretrievably from reality. Perhaps one of the most frequently used and poorly rationalized concepts relates to adhesion between the contacting asperities. The nature of the assumed adhesion is never described, nor is evidence for adhesion shown.

Nature of supporting information, especially from experiments. The most helpful papers include data from experiments covering a wide range of variables such as sliding speed, surface roughness, etc. Furthermore, papers that include a lengthy analysis of previous work are much more likely to be placing the new results in the proper context than are those in which references are merely cited. Finally, those papers in which wear rates and transitions in wear rates are identified with observations (with and without microscopes) of the surface's appearance, nature of wear particles, and other important features of surfaces are more useful than others.

Results of Applying the Above Criteria to Equations in Erosion

Erosion by solid particle impingement is one of the less complicated types of wear. Ninety-eight equations were found in this topic, but not all of them survived careful scrutiny. When the above four criteria were applied, 28 equations appeared useful for further analysis. Few of these equations contain the same array of factors: 28 equations contained some 33 factors, counting a few combined and adjustable coefficients. These are shown in Figure 10.2.

It is interesting to speculate on the reasons authors choose factors. Academic specialty is clearly one reason. There is really no sure way to discern whether all of the necessary variables and parameters were included or only the convenient ones. Dimensional analysis has been used by some authors to choose factors, but inherent in this method is the assumption that the only relevant factors are those that happen to have useful units.

One logical indicator of completeness in erosion equations might be the exponent on velocity, V. These range from about 1.5 to 6. Ordinarily one would expect that this exponent should be 2, reflecting the idea that particle energy would be the operative measure of impact severity. Several authors note that the larger exponents are found when ceramic materials are the target. Perhaps variation in this exponent actually reflects the response of the target materials to the strain rate differences inherent in differences in V. Thus an exponent of other than 2 may indicate that dynamic material properties should be used in equations rather than static properties. Alternatively an exponent other than 2 may indicate the absence of one or more important variables. To pursue these possibilities it is necessary to acquire a large body of data from tests using wide ranges of the many variables.

Equation #	1	2	3	4	5	6	7	8	9	10	11	12	13	14	15	16	17	18	19	20	21	22	23	24	25	26	27	28
Particle																												
Density						×			×		×				×	×	×	×				×	×	×			×	×
Hardness						×	×				×			×														
Moment of Inertia								×																				
Roundness						×	×				×			×														
Single mass								×						×														
Size			×				×	×	×	×			×		×	×	×	×	×	×							×	×
Velocity	×	×	×	×	×	×	×	×	×	×	×	×	×		×	×	×	×	×	×	×	×	×				×	×
Rebound velocity				×																								
Particle kinetic energy																									×			
Target																												
Density	×	×	×	×		×	×	×	×		×			×		×	×	×			×	×	×	×	×		×	×
Hardness						×	×		×		×				×	×	×	×			×	×	×	×			×	
Flow stress								×													×							
Young's Modulus																									×		×	
Fracture toughness																×	×	×						×	×		×	×
Critical strain/impact																					×					×		
Depth of penetration																										×		
Incremental strain																										×		

Figure 10.2 Chart showing which parameters were used in 28 models for solid particle erosion.

Equation #	1	2	3	4	5	6	7	8	9	10	11	12	13	14	15	16	17	18	19	20	21	22	23	24	25	26	27	28
Thermal conductivity														×														
Melting temperature														×								×						
Enthalpy of melting														×														
Cutting energy		×		×																								
Deformation energy		×		×																								
Erosion resistance						×	×				×																	
Heat capacity																						×	×					
Gram molecular wt														×														
Weibull flaw parameter			×				×																					
Time constant																	×	×										
Grain diameter																									×			
General																												
Impact angle	×	×	×	×		×	×	×			×	×	×						×	×	×	×						
Impact angle @max. loss												×																×
KE transfer, P-S														×														
Temperature																						×						
Prop. constants, # of	2	3	3	1	2	10	3	2	1	6	1	4	4	1	1	1	1	1	1	4	1	8	1	1	1	1	3	3

Figure 10.2 (continued).

Observations

The primitive state of equation development in friction, scuffing, and wear is clearly the result of the complexity of the topics. This complexity is obvious to anyone who sets out to understand the field, but there are two indications of this complexity in the research methods reported in the literature. These are:

1. The efforts of the several disciplines in the field are poorly coordinated.
2. There is insufficient consensus on coherent methods of constructing equations.

The efforts of the several disciplines in the field are poorly coordinated. This is apparent when classifying the topics of papers along the continuum of stages in the development of equations.

The developments in deriving equations of friction wear may be compared with parallel developments in deriving equations of all phenomena. These activities are often very chaotic, alternating between periods of high and low activity. Seldom is there identifiable orchestration of the exercise, and advances appear from seemingly random locations. In retrospect it can usually be seen that certain cultural, technical, economic, and even climate-related circumstances influence (for better or worse) the progress in understanding various phenomena.

Generally there are certain identifiable steps toward the goal of modeling. These steps include the following, called stages and expressed in terms of friction and wear.

STAGE A. The conscious and directed search for the variables that influence friction and wear. This may be compared with an exercise in scouting, as when routes were sought for crossing the Rocky Mountains in wagons to get to the west coast of the US. Many people in the east had some knowledge of the terrain to their very local and immediate west, but few people were able to assemble a total impression of the mountains. Some people, particularly in the military, devoted a great amount of time to gaining that impression, but it was done by covering the terrain in person, in the face of considerable hardship.

STAGE B. The summarizing and exchange of experience, as in publications, discussions at conferences, etc. These, like the accounts of scouts, are not often accurate, are biased according to the expertise of the reporter, and not perfectly communicated. Strong personalities prevail over the more modest reporters and the steady performers are not always heard.

STAGE C. A slow sifting of claims and a gaining of confidence in certain reporters who regularly support their claim with convincing evidence.

STAGE D. A bold stroke in some small area, perhaps done by some less well-known but attentive individual, who presents an equation or other broadly stated concept that feels right to many in the audience.

STAGE E. The slow adopting of the new ideas and abandonment of unprofitable lines of thinking in the research community.

STAGE F. Widespread use of the new ideas in engineering practice.

These stages may be identified in the field of wear as well, but different parts of knowledge in friction and wear are in different stages of development along the pathway from A to F. Overall, we find that researchers in materials sciences tend to focus on Stages A and B, without moving very resolutely toward later stages. Researchers in mechanics tend to focus on Stage D but do so without sufficient scouting or sifting of the knowledge derived from close studies of the basic mechanisms of friction and wear.

There is insufficient consensus on coherent methods of constructing the necessary equations. This may be seen in the continued use of a limited set of material parameters in particular and omission of other important ones. For example:

1. Few of the mechanical properties used in tribology equations are unique, i.e., many of them are the result of the same basic behavior of atoms, e.g., hardness, Young's Modulus, and melting point, and yet several are found together in most equations.
2. Some of the properties are not intrinsic material properties, such as hardness or stress intensity factors.
3. Some variables should not be found in first-principles wear equations, such as temperature or the coefficient of friction. Temperature does not cause wear, but does influence the material properties that control wear (among other things). The basic mechanisms of friction and wear are probably the same, so one cannot be used to describe the other.
4. Few of the properties are related to the mechanisms whereby wear particles are generated.
5. Some properties are rarely seen in equations though sliding of one surface over another certainly calls for these properties, e.g., fatigue properties, oxide properties, debris content and stickiness, strain rate sensitive mechanical properties, etc.
6. The influence of geometry and other factors on wear particle retention is not often considered.
7. The influence of vibration, slight deviation from repeat-pass paths, the difference between cyclic sliding, repeat-pass and single-pass sliding are not usually considered.
8. Most researchers select variables for study exclusively from their own discipline.
9. Wearing usually takes place by a combination of mechanisms, which changes with time. There are few studies on transitions between the balance of mechanisms and on partitioning between mechanisms.

In the absence of good equations designers must use other methods to select materials and safe operating conditions. Some methods are recommended in Chapter 11.

REFERENCES

1. Meng, S.H., PhD research, Mechanical Engineering Department, University of Michigan, Ann Arbor, 1994.
2. Lim, S.C., Ashby, M.F., and Brunton, J.H., *Acta Met*, 35, 1343, 1987.

CHAPTER 11

Designing for Wear Life and Frictional Performance: Wear Testing, Friction Testing, and Simulation

SINCE THERE ARE NO GENERALLY USEFUL EQUATIONS FOR FRICTION, SCUFFING, OR WEAR, DESIGNERS MUST PROCEED BY COMBINATIONS OF TESTING, CONSULTATION WITH PEOPLE ASSOCIATED WITH SIMILAR PRODUCTS, AND GOOD DOCUMENTATION FROM DEVELOPMENT PROJECTS.

INTRODUCTION

Warranty costs due to unpredicted wear and undesirable frictional behavior (coefficient, vibration tendency, stability, etc.) exceed that for most other mechanical causes *combined* in some industries. Effective methodologies for good tribological design can be very cost effective.

The wear life and frictional stability of mechanical components involves nearly as many variables as those that affect human life. Sliding surfaces are composed of substrate material, oxide, adsorbed substances, and dirt. They respond to their environment, method of manufacture, and conditions of operations. They suffer acute and/or progressive degeneration, and they can often be partially rehabilitated by either a change in operating conditions or by some intrusive action.

DESIGN PHILOSOPHY

Most sliding surfaces are redesigned rather than designed for the first time. Thus a designer will usually have access to people who have experience with previous products. Designing a product for the first time requires very mature skills, not only in materials and manufacturing methods but also in design philosophy for a particular product.

The philosophy by which the wear life or frictional behavior of a product is chosen may differ strongly within and between various segments of industry. Such considerations as acceptable modes of failure, product repair, controllability of environment, product cost, nature of product users, and the interaction between these factors receive different treatment for different products. For example, since automobile tires are easier to change than an engine crank, the wear life of tires is not a factor in a discussion of vehicle life. The opposite philosophy must apply to drilling bits used in the oil industry: the cone teeth and the bearing upon which the cone rotates must be designed for equal life since both are equally inaccessible while wearing.

In some products or machines, function is far more important than manufacturing costs. One example is the sliding elements in nuclear reactors. The temperature environment of the nuclear reactor is moderate, lubricants are not permitted, and the result of wear is exceedingly detrimental to function of the system: expensive metal–ceramic coatings are frequently used. This is an example of a highly specified combination of materials and wearing conditions. Perhaps a more complex example is that of artificial teeth. The surrounding system is very adaptable, a high cost is relatively acceptable, but durability may be strongly influenced by body chemistry and choice of food, all beyond the designer's range of influence.

Thus, there is no general rule whereby a designer can quickly proceed to select an acceptable sliding material for a product. One oft-heard but misleadingly simple method of reducing wear is to increase the hardness of the material. There are, unfortunately, too many exceptions to this rule to have high confidence in it except for some narrowly defined wearing system. One obvious exception is the case of bronzes, which are more successful as a gear material against a hardened steel pinion than is a hardened steel gear. The reason usually given for the success of bronze is that dirt particles are readily embedded into the bronze and therefore do not cut or wear the steel away, but this is more of an intuitive argument than fact. Another exception to the hardness rule is the cam in automotive engines. They are hardened to the range of 50 Rockwell 'C' instead of to the maximum available, which may be as high as 67 Rc. A final example is that of buckets and chutes for handling some ores. Rubber is sometimes found to be superior to very hard white cast iron in these applications.

We see in the examples above the possibility of special circumstances requiring special materials. The rubber offers resilience, and the cam material resists fatigue failure if it is not fully hardened. It is often argued that special circumstances are rare, or can be dealt with on a case-by-case basis. This attitude seems to imply that most sliding systems are standard, thus giving impetus to specifying a basic wear resistance or friction of material as one of its intrinsic properties. Little real progress has been made in this effort and very little is likely to be made in the near future. Wear resistance and frictional behavior are achieved by a balance of several very separate properties, not all of them intrinsic, and different for each machine component or sliding surface. Selecting the best material for wear resistance is therefore a complex task and guidelines are needed in design.

Guidelines will be more useful as our technology progresses, and some are given below.

STEPS IN DESIGNING FOR WEAR LIFE OR FRICTIONAL BEHAVIOR WITHOUT SELECTING MATERIALS

The Search for Standard Components

Designers make most of the decisions concerning material selection. Fortunately, for many cases and for most designers, the crucial components in a machine in which wear or friction may limit useful machine function are available as separate packages with fairly well-specified performance capabilities. Examples are gear boxes, clutches, and bearings. Most such components have been well tested in the marketplace, having been designed and developed by very experienced designers. For component designers, very general rules for selecting materials are of little value. They must build devices with a predicted performance of ± 10 percent accuracy or better. They know the range of capability of lubricants; they know the reasonable range of temperature in which their products will survive; and they know how to classify shock loads and other real operating conditions. Their specific expertise is not available to the general designer except in the form of the shapes and dimensions of hardware, the materials selected, and the recommended practices for use of their product. Some of these selections are based on tradition, and some on sound reasoning strongly tempered by experience. The makers of specialized components usually also have the facilities to test new designs and materials extensively before risking their product in real use. General designers, on the other hand, must often proceed without extensive testing.

The general designer must then decide whether to use standard, specialized components or whether to risk designing every part personally. Sometimes the choice is based on economics, and sometimes desired standard components are not available. In such cases components as well as other machine parts must be designed in-house.

In-House Design

It is logical for designers to follow the methods used in parallel activities such as in determining the strength and vibration characteristics of new machinery. This is often done by interpolating within, or extrapolating beyond, known experience, if any, using four sources:

Company practice for similar items: If good information is available on similar items, a prediction of the wear life of a new product can be made within ±20 percent accuracy unless the operating conditions of the new design are much beyond standard experience. Simple scaling of sizes and loads is often successful, but usually this technique fails after a few iterations. Careless comparison of a new design with similar existing items can produce very large errors for reasons discussed below.

When a new product must be designed that involves loads, stresses, or speeds beyond those previously experienced it is often helpful to review the recent performance and examine in detail the worn surface of a well-used previous model. It is also helpful to examine unsuccessful prototypes or friction/wear test specimens as will be discussed below. An assessment should be made of the modes or mechanisms of surface change of each part of the product. For this purpose it is also useful to examine old lubricants, the contents of the lubricant sump, and other accumulations of residue.

Vendors of materials, lubricants, and components: When a new product requires bearings or materials of higher capacity than now in use, it is frequently helpful to contact vendors of such products. Where a vendor simply suggests an existing item or material from his brochure, the wear life of a new product may not be predictable to an accuracy of better than 50 percent of that desired. This accuracy is worse than the ±20 percent accuracy given above especially where there is inadequate communication between the designer and the vendor. Accuracy may be improved when an interested vendor carefully assesses the needs of a design, supplies a sample for testing, and follows the design activity to the end.

Contact with vendors, incidentally, often has a general beneficial effect. It encourages a designer to explore new ideas beyond the simple extrapolation of previous experience. Most designers need a steady flow of information from vendors to remain informed on both the new products and on the changing capability of products.

Handbooks: There are very many handbooks, but few that assist substantially in selecting materials for wear resistance or frictional behavior. Materials and design handbooks usually provide lists of materials some of which are highlighted as having been successfully used in sliding parts of various products. They usually provide little information on the rates of wear of products, the mode of wear failure, the variations of friction, the limit on operating conditions, or the method by which the sliding parts should be manufactured or run-in (if necessary).

Many sources provide tables of coefficient of friction as mentioned in Chapter 6, and some sources will give wear coefficients which are purported to be figures of merit or ranking of materials for wear resistance. A major limitation of friction and wear coefficients of materials as given in most literature is that there is seldom adequate information given on how the data were obtained. Usually this information is taken from standard laboratory bench tests, few of which simulate real systems, and few of which rank (order) materials for wear life in the same way that production parts rank in the hands of the consumer. The final result of the use of handbook data is a design which will probably not perform to an accuracy of better than ±75 percent.

Equations: Wear is very complicated, involving up to seven basic mechanisms, operative in different balance or ratio in various conditions, and many of the mechanisms produce wear rates that are not linear in the simple parameters such as applied load, sliding speed, surface finish, etc. There are, at this time, no complete first principles or models available to use in selecting materials for wear resistance. (See Chapter 10.) However, there are good procedures to follow in selecting material for wear resistance.

STEPS IN SELECTING MATERIALS FOR WEAR RESISTANCE

When designing for wear resistance, it is necessary to ascertain that wear will proceed by the same combination of mechanisms throughout a substantial portion of the life of a product: only then is reasonable prediction of life possible.

Following are those considerations that are vital in selecting useful materials, and they may be more important than selecting the most wear-resisting material.

Determine whether there are restrictions on material use: In some industries it is necessary for economic and other purposes to use, for example, a gray cast iron, or a material that is compatible with the human body, or a material with no cobalt in it such as in a nuclear reactor, or a material with high friction, or a selected surface treatment applied to a low-cost substrate. Furthermore, there may be a limitation on the surface finish available, or the skill of the personnel who will manufacture or assemble the product. Finally, there may be considerations of delivery, storage of the item before use, disposal after use, or several other events that may befall a wear surface.

Determine whether the sliding surface can withstand the expected static load without indentation or excessive distortion: Generally, this would involve a simple stress analysis.

Determine the sliding severity that the materials must withstand in service: Factors involved in determining sliding severity include the contact pressure or stress, the temperature due to ambient heating and frictional temperature rise, the sliding speed, misalignment, duty cycle, and type of maintenance the designed item will receive. Each factor is explained below.

a. *Contact stress*: Industrial standards for allowable contact pressure vary considerably. Some specifications in the gear and sleeve bearing industries limit the average contact pressures for bronzes to about 1.7 MPa, which is about 1 percent to 4 percent of the yield strength of bronze. Likewise, in pump parts and valves made of tool steel the contact pressures are limited to about 140 MPa which is about 4 to 6 percent of the yield strength of the hardest state of tool steel.

   One example of high contact pressure is the sleeve bearings in the landing gear of one commercial airplane, the DC9. These materials again are bronzes and have yield strengths up to 760 MPa. The design bearing stress is 415 MPa but with expectations of peak stressing up to 620 MPa. Another example is the use of tool steel in lubricated sheet metal drawing. Dies may be used for 500,000 parts with contact pressures of about 860 MPa, which is half the yield strength of the die steel.
b. *Temperature* strongly influences the life of some sliding systems. Handbooks often specify a material for wear conditions without stating a range of temperature within which the wear resistance behavior is satisfactory. The influence of temperature may be its effect on the mechanical properties of the sliding parts. High temperatures soften most materials and low temperature embrittles

some. High temperature will produce degradation of most lubricants but low temperature will solidify a liquid lubricant.

Ambient temperature is often easy to measure, but the temperature rise due to sliding may have a larger influence (see Figure 5.15 of Chapter 5). Thermal conductivity of material could be influential in controlling temperature rise in some cases, but a more important factor is, μ, the coefficient of friction. If a temperature sensitive wear mechanism is operative in a particular case then high friction may contribute to a high wear rate, if not cause it. There is at least a quantitative connection between wear rate and μ when one compares dry sliding with adequately lubricated sliding, but there is no formal way to connect μ with ΔT.

c. *Sliding speed and the PV limits*: Maximum allowable loads and sliding speeds for materials are often specified in catalogs in the form of PV limits. (PV limits are discussed in Chapter 8.) A PV limit indicates nothing about the actual rate of wear of materials, only that above a given PV limit a very severe form of wear may occur.

d. *Misalignment*: Where some misalignment may exist it is best to use a material that can adjust or accommodate itself, i.e., break in properly. Misalignment arises from manufacturing errors from a deflection of the system producing loading at one edge of the bearing, or it may arise from thermal distortion of the system, etc. Thus, a designer must consider designing a system such that a load acts at the expected location in a bearing under all conditions. This may involve designing a flexible bearing mount, or several bearings along the length of a shaft, or a distribution of the applied loading, etc.

A designer must also consider the method of assembly of a device. A perfectly manufactured set of parts can be inappropriately or improperly assembled, producing misalignment or distortion. A simple tapping of a ball bearing with a hammer to seat the race may constitute more severe service than occurs in the lifetime of the machine and often results in early failure.

Misalignment may result from wear. If abrasive species can enter a bearing the fastest wear will occur at the point of entry of the dirt. In that region, the bearing will wear away and transfer the load to other locations. A successful design must account for such events.

e. *Duty cycle*: Important factors in selecting materials for wear resistance are the extent of shock loading of sliding systems, stop-start operations, oscillatory operations, etc. It is often useful to determine also what materials surround the sliding system, such as chemical or abrasive particles.

f. *Maintenance*: A major consideration that may be classified under sliding severity is maintenance. The difference between industrial and aircraft use includes different treatment of bearings in maintenance. Industrial goals are to place an object into service and virtually ignore it, or provide infrequently scheduled maintenance. Aircraft maintenance, on the other hand, is more rigorous and each operating part is under regular scrutiny by the flight crew and ground crew. Thus, it is easier for an error to be made in selection of lubricant in industry than with aircraft, for example. Further, the aircraft wheel bearing operates in a much more standard or narrowly defined environment. Industrial machinery must operate in the dirtiest and hottest of places with the poorest care.

Determine whether or not a break-in procedure is necessary or prohibited: It cannot be assumed that the sliding surfaces made to a dimensional accuracy and specified surface finish are ready for service. Sliding alters surfaces. Frequently, sliding under controlled, light loads can prepare a surface for a long life of high loading, whereas immediate operation at moderate loads may cause early failure.

It is useful here to distinguish between two surface-altering strategies. The first is where a system is immediately loaded or operated to its design load. The incidence of failure of a population of such parts decreases with time of operation as the sliding surfaces change, and frequently the ability of the system to accommodate an overload or inadequate lubricant increases in the same time. The surfaces have changed in some way during running. Systems may also be operated in a deliberate and planned manner that prepares them for normal service. This latter process was referred to as "break-in" in Chapter 9.

The wear that occurs during break-in can be considered a final modification to the machine surface. This leads to the possibility that a more careful specification of manufacturing practice may obviate the need for run-in or break-in. Only 60 years ago it was necessary to start and run an engine carefully for the first few thousand miles to ensure a reasonable engine life. If run-in were necessary today one would not see an engine survive the short trip from the assembly plant to the haul-away trucks!

It is difficult to determine if some of the present conservative industrial design practices result from the impracticality of breaking-in some products. For example, a gear box on a production machine is expected to function immediately without break-in. If it were broken in, its capacity might be greatly increased, but for each expected severity of operation of a device, a different break-in procedure is necessary. Thus, a machine that has been operating at one level of severity may be no more prepared for a different state of severity than if it had never been run. A safe procedure, therefore, is to operate a device below the severity level at which break-in is necessary, which really amounts to overdesigning.

Determine acceptable modes of wear failure, surface damage, or debris form: To specify a wear life in terms of a rate of loss of material is not sufficient. For example, when an automotive engine seizes up, there is virtually no loss of material, only a rearrangement such that function is severely compromised. In some machines, surface rearrangement or change in surface finish is less acceptable than attrition or loss of material from the system. In metal working dies, loss of material from the system is less catastrophic than is scratching of the product. Finally, in some systems, particularly in artificial human joints and computer hard disks, the wear debris is a greater hazard than is a loss of dimension from the sliding members.

In truck brakes some abrasiveness of brake linings is desirable even though it wears brake drums away because that wear removes microcracks and avoids complete thermal fatigue cracking. On the other hand, in cutting tools, ore crushing equipment, and amalgam fillings in teeth, surface rearrangement is of little consequence, but material loss is to be avoided.

A final example of designing for an acceptable wear failure are the sleeve bearings in engines. Normally they should be designed against surface fatigue.

However, in some applications corrosive conditions may seriously accelerate fatigue failure. This may require the selection of a material that is less resistant to dry fatigue than is the best bearing material, and this applies especially to the two-layer bearing materials. In all of these examples a study of acceptable modes of wear may result in a different selection of material than if the goal is simply to minimize wear.

Decide whether or not to begin wear testing: After some study of worn parts from a device or machine that most nearly approximates the new or improved product, one of several conclusions could be reached.

a. The same design and materials in the wearing parts of the example device will perform adequately in the redesign, both as to function, cost, and all other attributes.
b. A slight change in size, lubrication, or cooling of the example parts will be adequate for the design.
c. A significant change in size, lubrication, or cooling of the example parts will be necessary for the redesign.
d. A different material will be needed in the redesign.

The action to be taken after reaching one of the above conclusions will vary. The first conclusion above can reasonably be followed by production of a few copies of the redesign. These should be tested and minor adjustments made to ensure adequate product life.

The second conclusion should be followed by cautious action, and the third conclusion should involve the building and exhaustive testing of a prototype of the redesign. The fourth conclusion may require tests in bench test devices, in conjunction with prototypes.

TESTING AND SIMULATION

It is costly and fruitless to purchase bench test machinery and launch into testing of materials or lubricants without experience and preparation. It is doubly futile for the novice to run accelerated wear tests, with either bench tests, prototypes, or production parts. Furthermore, an engineer learns very little by having wear tests done by a distant technician who supplies the engineer with cleaned up specimens and data on equilibrium wear rate at the end of the test.

The problem is that *wear resistance is not a single property of any material.* Hardness, Young's Modulus, and density are single properties which may be measured in standard tests. Wearing of material, by contrast, occurs by a succession of changing balances of mechanisms, controlled by the sliding situation. To use a familiar analogy, material attributes, such as hardness, are equivalent to the height, weight, and eye color of humans, whereas the expected life of a person is determined by many mechanisms, including bodily condition, lifestyle, and external events. Generally, we do not expect to predict life expectancy by measuring body temperature alone; most people consider other known factors, and we usually consult physicians for a reasonably precise prediction of remaining human life.

Testing of tribological systems can provide some information on likely product life. The tests must be well designed but the proof of appropriateness of a test depends more on the test results than on how the test is done. Types of test, test parameters, and other details are discussed in the following paragraphs, followed by a suggested criterion for degree of correlation or simulation between wear test results and wearing of the design under study.

Standard Tests and Test Devices

Standard test devices are described in several references.[1] Some of them were developed as wear testers, such as the dry sand–rubber wheel test. Many others were developed for testing lubricants, such as the 4-ball tester. A few have been developed to measure friction, such as the (tire) skid resistance machine.

Most test devices simply slide two specimens together in a simple manner whereas a few, such as a hip joint simulator attempts to emulate the motion of the implanted specimen during walking or running.

Many, if not most, available test devices are named in the standards of the Society of Automotive Engineers (SAE), the American Society of Standards and Materials (ASTM), and the standards-making societies of other nations. Many standards are standard *methods* for operating the test devices so that everyone who uses identical devices will obtain very nearly identical results. This is a useful exercise where products, materials, and lubricants are to be compared or ranked for some quality. However, the measured or inferred quality may not be the quality connected with assured product life. In short, the standard test may not (probably does not) simulate the experience of material or lubricants in practical machinery. Methods for determining how well test devices and test procedures simulate practical machinery will be discussed later in this chapter. Clearly, no standard test method assures simulation with any real product.

A clear indication of the problem with bench tests may be seen in some results with three test devices. These are:

1. Pin-V test in which a 1/4-inch-diameter pin of 3135 steel rotates at 200 rpm with 4-line contact provided by two V blocks made of 1137 steel.
2. Block-on-ring test where a rectangular block of a chosen steel slides on the outer (OD) surface of a ring of hard, case carburized 4620 steel.
3. The 4-ball test where a ball rotates in contact with three stationary balls, all of hardened 52100 steel.

The 4-ball test and the block-on-ring test were run over a range of applied load and speed. The pin-V test was run over a range load only. All were run continuously, i.e., not an oscillating or stop-start sequence mode. All tests were run with several lubricants.

Results from the block-ring test were not sufficiently reproducible or consistent for reliable analysis. Results from the other two tests were adequate for the formulation of a wear equation from each, as follows:

$$\text{Pin-V test: Wear rate} \propto (\text{Load})^2$$

$$\text{4-ball: Wear rate} \propto (\text{Load})^{4.75} \times (\text{Speed})^{2.5}$$

These results may be compared with linear laws of wear discussed frequently in the literature, which would be of the form:

$$\text{Linear law: Wear rate} \propto (\text{Load})^{1.0} \times (\text{Speed})^{1.0}$$

Necessary Variables to Consider in Wear Testing

The number of necessary variables that influence wear rate probably exceeds 75. Some variables are not usually explicitly mentioned but are embodied in the choice to use the same materials, prepared in the same way as that in the practical sliding pair under study.

Perhaps the single and most important error committed in wear testing is to assume that an adequate test is one in which the contact pressure (load) and sliding speed is the same as in the practical device under design. If this were the case, the equations embodying the test results given above would have identical exponents. The fact that they do not, most likely indicates that the results are sensitive to several of the unmentioned test parameters or material (including lubricants) variables. There are no methods for determining which parameters and variables are missing as discussed in Chapter 10. There is no exhaustive list of test variables available, but for illustration a few are mentioned briefly:

1. Contact shape: A sphere-on-flat test versus a cylinder-on-flat (Figure 9.3 of Chapter 9) produce different results, probably because wear particles are recycled differently in the two tests. Particles are swept aside in the sphere-on-flat test but are constrained to pass through the contact region in the cylinder-on-flat test. The result is a great difference in load-carrying capacity in boundary lubricated sliding.
2. System vibration: The smallest effect of vibration is a time-varying contact stress, the most significant is the alteration of wear particle movement.
3. Tracking variability in repeat pass sliding produces varying results.
4. Reciprocating sliding accommodates wear particles and fluid films differently than do circular repeat pass sliding.
5. Oxygen availability to the contact region influences the types and amounts of oxides that form.
6. Duty cycle and standing time influence temperature and surface chemistry.

Accelerated Tests

The most common way of conducting an accelerated test is to increase either the load or sliding speed. The hazard of such procedures may be seen in two ways. The first is by considering the equations discussed under *Standard Tests and Test Devices* above. The influence of load and speed is different in each test, and certain to be different again in the device under design study. An accelerated test has no meaning if the influence of accelerated conditions is not known.

A second point is made by examining Figure 8.11 in Chapter 8. If the design under study (using steel of 348 VPN hardness) is to operate with a load of 20 grams, an accelerated test in which 200 grams of load is applied would indicate an unacceptable wear rate. Conversely if the design uses a load of 200 grams, a test where 2000 grams is applied would produce deceiving results. It should be noted that the curve for the wear rate of 348 VPN steel can be characterized by a much more complicated equation than those given in *Standard Tests and Test Devices*.

Criterion for Adequate Simulation

Experience shows time after time that *simple* wear tests complicate the prediction of product life. The problem is correlation or assurance of simulation. For example, automotive company engineers have repeatedly found that engines on dynamometers must be run in a completely unpredictable manner to achieve the same type of wear as seen in engines of cars in suburban use. Engines turned by electric motors, though heated, wear very differently from fired engines. Separate components such as a valve train can be made to wear in a separate test rig nearly the same way as in a fired engine, with some effort, but cam materials rubbing against valve lifter materials in a bench test inevitably produce very different results from those in a valve train test rig.

Most machines and products are simpler than engines, but *the principle of wear testing is the same, namely, the combination of wear mechanisms must be very similar in each of the production designs, the prototype test, the subcomponent test, and the bench test. The wear rates of each test in the hierarchy should be similar, the worn surfaces must be nearly identical, and the transferred and loose wear debris must contain the same range of particle sizes, shapes, and composition.* Thus it is seen that the prototype, subcomponent, and bench tests must be designed to correlate with the wear results of the final product. It would be best also if the measured coefficient of friction, contact resistance, and approximate surface temperature were also similar. This requires considerable experience and confidence when the final product is not yet available. This is the reason for studying the worn parts of the product nearest to the redesign and a good reason for retaining resident wear expertise in every engineering group.

Measurements of Wear and Wear Coefficients and Test Duration

The measurements of material loss to be taken from a test could take any form that can eventually be used to predict the end of the wear life of a product. These include:

- Wear volume/unit distance of sliding
- Mass loss/unit of time
- Change in wear track width/unit distance of sliding, etc.

Conversion from one form to another may require knowledge of material density.

The measurement of mass loss often requires longer time tests than other methods, which can cost more money than necessary. Generally, it is useful to devise ways to make precise measurements of volume loss than to wait for sufficient mass loss to measure. One precise method is to use a surface tracer system to measure the profile of the wear scar. With a properly programmed computer, several parallel traces can be converted into a volume of material loss. An important caveat in using precise methods is to assure first in a test series that wear is progressing in the same way early in a test as it would late in a test.

MATERIAL SELECTION TABLE

Selecting materials for wear resistance requires a study of the *details* of wear in a wearing system (including the solids, the lubricant, and all of the wear debris) such as an old product being redesigned or a wear tester. The tools for such examination are described in Chapter 12. The designer can then proceed through the following table and make a first attempt at selecting material for wear resistance. The table is used in the following manner:

1. Observe the nature of worn surfaces and debris in existing devices, or of similar materials in appropriate (closely simulating) wear testing machines.
   NOTE: Observations should be done with appropriate devices and instruments and at the proper scale.
2. Check the lists in Section A below for an applicable description of worn surfaces or type of service, noting the code that follows the chosen term.
3. Proceed to Section B and verify that the code listing is an adequate description of the worn surface. (It is possible to use Section B without reference to Section A.) From Section B find the major term (CAPITALIZED) in Columns a, b, c and d. Columns e and f are added to complete the description of the surface.
4. Find the definition or detailed description of the (CAPITALIZED) major term in Section C, and note which MATERIAL LOSS MECHANISM is applicable, and confirm that the nature or description of wear debris is consistent with the chosen wear mode.
5. Find the MATERIAL LOSS MECHANISM again in Section D and note the material characteristics and microstructure that should influence wear resistance

of material, and note the precautions in material selection to prevent material failure.

6. Select materials in conjunction with materials specialists. It is helpful to know that materials specialists are able to specify a material for wear resistance best with complete knowledge of modes of wear of the proposed design, and that the first choice may not be successful, either functionally or economically.

(See Problem Set questions 11 a and b.)

SECTION A. COMMON EXPRESSIONS FOR TYPES OF WEAR (WEAR INCLUDES MATERIAL LOSS AND SURFACE DAMAGE)

EXPRESSIONS CONNECTED WITH APPEARANCE OF SURFACES	EXPRESSIONS CONNECTED WITH TYPE OF SERVICE	
Stained -f	Surface corrosion or	In solid machinery -a1+c
Polished, or smooth wear -a1+c+e or a2+c+e	Erosion-corrosion	In fluids - a2+d2
Scratched (short grooves) - b3+c+e	Abrasive wear - b3+c	(Multiple scratches)
Gouged - b3+d1		
Scuffed-a1+initiated and periodically perpetuated by d3, +e	Gouging - b1+d1+e	
Galled - b1+d3+e (usually very rough)	Dry wear or unlubricated sliding - b1+d3+e, or a1+c+e	
Grooved (smooth or rough -a1+periodically advanced by d1 +e	Metal-to-metal wear, or adhesive wear - b1+d3+e	
Hazy - b2	Erosion at high angle - b2+d4	
Exfoliated or delaminated - d4+e	Erosion at low angle - b3+d1 or d2	
Pitted - b2 and/or d5		
Spalled - d4		
Melted - a3+?		
Fretted - a1+d5+f	Fretting - a1+d5+f	

Rigorous connection cannot always be made between the terms in the two columns because of wide diversity of use and meaning of terms.

SECTION B. LIST OF SURFACE PHYSICAL CHARACTERISTICS AND THE PROCESSES THAT PRODUCE THEM

a - micro-smooth	b - micro-rough
1. Progressive loss and reformation of surface films, e.g., oxide (oxidative wear?), others (erosion-corrosion?), by fine ABRASION and/or tractive stresses, mutually imposed by ADHESIVE or viscous interaction 2. Very fine ABRASION, with loss of substrate in addition to loss of surface film, if any 3. From MELTING	1. Due to tractive stresses resulting from ADHESION 2. Micro-pitting by FATIGUE 3. ABRASION by medium-coarse particles
c - macro-smooth	**d - macro-rough**
1. ABRASION by fine abrasives held on solid backing (lapping, polishing) (usually removing only oxides)	1. ABRASION by coarse particles, including carbide and other hard inclusions in the sliding materials,which are removed by sliding action as wear of matrix progresses 2. ABRASION by fine particles in turbulent fluid, producing scallops,waves, etc. 3. Severe ADHESION, at least as an initiator of damage 4. Local FATIGUE failure resulting in pits or depressions, repeated rolling contact stress, or repeated thermal gradients, or repeated high friction sliding, or repeated impact by hard particles as in erosion 5. Advanced stages of micro-roughening,
e - shiny	**f - dull or matte**
Very thin (or perhaps no surface film) of e.g., oxide, hydroxide, sulfide, chloride, or other species	Thick films of perhaps greater than 25nm thickness (resulting from "aggressive environments" including high temperature)

Careful observation usually reveals at least two scales of events, micro- and macro-, (omitting the several submicroscopic events that are known to occur).

SECTION C. DEFINITION OF MAJOR TERMS, INCLUDING THE NATURE OF DEBRIS AND MOST PROBABLE WEAR PROCESS, (THE MOST LIKELY MECHANISM OF MATERIAL LOSS IS CAPITALIZED AND UNDERLINED)

	Nature of Debris
CORROSION (of surfaces in this case) — chemical combination of material surface atoms with passing or deposited active species to form a new compound, original, i.e., oxide, etc.	The newly formed chemical compound, usually agglomerated, sometimes mixed with fragments of the surface material
Abrasion — involves particles that have some acute angular shapes but are made mostly of obtuse shapes. These form wear debris. Some debris forms ahead of the abrasive particle. This is called *CUTTING*. However, most debris is material that has been plowed aside repeatedly by passing particles and breaks off in long, often curly, chips or strings by *LOW CYCLE FATIGUE.*	
Adhesion — a strong bond that develops between two surfaces, either between coatings and/or substrate materials, which with relative motion produced a tractive stress that may be sufficient to deform materials to fracture. The mode of fracture will depend on the property of the material, involving various amounts of energy loss, or ductility to fracture, i.e., low energy and ductility (BRITTLE FRACTURE) or high energy and ductility (DUCTILE FRACTURE).	Solid particles, often with cleavage surfaces (brittle fracture) Severely deformed solids, often with oxides mixed in (ductile fracture)
Fatigue — due to cyclic strains — usually at stress levels below the yield strength of the material, also called — *HIGH CYCLE FATIGUE.*	Solid particles, often with cleavage surfaces and ripple pattern
MELTING from very high-speed sliding.	Spheres, solid or hollow, and "splat'" particles

Debris may not indicate basic processes but is useful to indicate trends, new events, and progressions, and it sometimes reveals unexpected causes of wear.

SECTION D. MATERIAL CHARACTERISTICS THAT RESIST THE SEVEN MECHANISMS OF MATERIAL LOSS

Seven mechanisms of material loss	State of materials to resist wear	Precautions for selecting a material to resist material loss*
Formation and removal of products of CORROSION.	Reduce corrosiveness of surrounding region, increase corrosion resistance of metal by alloy addition, or select soft, homogeneous metal, or ceramics or polymers.	Total avoidance of new chemical species can result in high adhesion of contacting surfaces; and soft material tends to promote galling and seizure.
CUTTING (may actually not be a unique mode of material failure, but rather a type of fracture.)	Achieve high hardness, either throughout, by surface treatments, or by coatings. Add very hard particles or inclusions such as carbides, nitrides, (ceramics) etc.	All methods of increasing cutting resistance cause brittleness and lower fatigue resistance.
DUCTILE FRACTURE	High strength, achieved by any method other than by cold-working or by heat treatments that produce internal cracks or large and poorly bonded intermetallic compounds.	
BRITTLE FRACTURE	Minimize tensile residual stress. For cold temperature insure low temperature transition, temper all martensite, use deoxidized metal, avoid carbides as in pearlite, etc., and assure a good bond between fillers and matrix in composites to deflect cracks.	Soft materials will not fail in a brittle manner, and will not resist cutting very well.
LOW-CYCLE FATIGUE	Use homogeneous materials and high-strength materials that do not strain-soften. Avoid over-aged metals or other two-phase systems with poor adhesion between filler and matrix.	
HIGH-CYCLE FATIGUE	For steel and titanium, apply stresses less than half the tensile strength (however achieved), and for other metals to be cycled fewer than 10^8 times, allow stresses less than 1/4 the tensile strength (however achieved). Avoid retained austenite; select pearlite rather than plate structure; avoid poorly bonded second phases; avoid decarburization of surfaces; avoid platings with cracks; avoid tensile residual stress or form compressive residual stress by carburizing or nitriding.	Calculation of "contact stress" should include the influence of tractive stress
MELTING	Use material with high melting temperature and/or high thermal conductivity.	

* Metals of high hardness or strength usually have low corrosion resistance; ceramics are all prone to early fatigue failure; polymers creep under constant load; and all materials with multiple phases and multiple desirable properties are expensive.

CHAPTER 12

Diagnosing Tribological Problems

MECHANICAL ENGINEERS GENERALLY HAVE LITTLE CONFIDENCE IN SPECIFYING DETAILED STUDIES OF WORN OR SCUFFED SURFACES. MATERIALS ENGINEERS DO SUCH STUDIES WELL, BUT ARE NOT SUFFICIENTLY INVOLVED IN THE MECHANICAL ASPECTS OF DESIGN TO APPLY THEIR RESULTS EFFECTIVELY. THIS CHAPTER PROVIDES GUIDELINES FOR BOTH.

INTRODUCTION

Designers and product engineers find that the prediction of friction and wear life of mechanical components is the most confusing exercise in their technical career. The reason is simple. As stated implicitly or explicitly in previous chapters: there are no useful handbooks or equations that one may use to calculate any useful quantities in these topics. There are neither precise nor approximate methods for estimating friction or wear rate from first principles.

Problems of friction and wear usually appear in the following circumstances:

1. When an old product must be upgraded to function at higher load, higher speed, etc., or must be redesigned to be made at lower cost. The usual practice is to hope that the present materials will suffice, but eventually extrapolation fails.
2. When some products fail in the field. Some problems may have appeared in the development phase of the product, but eventually the problems seemed to have been solved. The product passed all tests and was released for production, but it was never totally trouble free.
3. An uncommon event, but sometimes a totally new product is to be made, and there is little or no in-house design to use as a guideline.

Most often, good design will come from well-informed designers. In the case of friction and wear, becoming well informed involves acquiring several personal skills besides becoming informed from the printed word. Acquiring tribological skills begins, however, with the internal conviction and confidence that there are no good guidelines in handbooks, in computer models, or any other place for designing products to meet specific friction or wear life requirements. Following are some specific useful skills:

1. Convince your management of the need for you to acquire tribological skills.
2. Read a good textbook all the way through and become determined to broaden well beyond your own disciplinary training.
3. Develop personal laboratory skills. In particular, learn how to identify types of wear by the behavior of the system in which the wearing parts are located. Observe the sliding surfaces in action, *feel* the vibration, examine the lubricant, the worn surfaces, and the wear debris.
4. Learn how to run the test machinery, observe how technicians conduct tests, and save all data.
5. Resist believing that data from laboratory test devices are useful until you verify some degree of correlation of lab data with the performance of real or prototype products.
6. When a problem becomes large in scope and involves organizational matters and relations with vendors, etc., learn how to use the team approach to solving tribological problems.

INTRODUCTION TO PROBLEM DIAGNOSIS

This section discusses methods of examining the surfaces of tribological systems. The systems are described as tribological systems in order to encompass problems that include wear by sliding, wear by erosion, chemically enhanced loss of material, friction without significant wear, and any other mechanical interaction of two substances, whether lubricated or not.

Surface examination may require the use of several types of instruments. Perhaps the most mysterious and beguiling are the chemical analysis instruments identified by the acronyms, ESCA, AES, LEED, EBS, etc. Whereas many such instruments are available, only a few will be described. Tribologists can usually solve two thirds of their problems with a small magnet, with low-power optical microscopy, and with surface roughness tracing. Few tribologists will need the more sophisticated instruments, and even fewer can be expected to know how to operate them. Five of the chemical analysis instruments will be described both so that you can determine whether you need them, and to gain a perspective on whether the many that are not described are worth looking into.

Little is written on methods of surface analysis for tribological problems. Analysis involves human decision as well as instruments. The best method of analyzing surfaces begins with a good plan, and the plan should include several steps. In the following it will be assumed that a problem is known to exist. Perhaps a candidate material is operating in some new device, and some judgment must be made as to its suitability. Perhaps some surfaces are wearing too quickly or in some undesirable pattern, or the surfaces may be sliding in some undesirable manner, and the time arrives to examine those surfaces. A procedure for surface analysis is given in the following paragraphs. The very first, and perhaps surprising, suggestion is to *avoid dismantling the device or cleaning the surfaces before you have devised a preliminary plan and conducted a fact-finding examination*, either formally or informally.

Planning

Assemble a group of people consisting of (depending on the size of the problem):

a. Engineers and technicians who have responsibility for the product under discussion. Wear is influenced by the system surrounding a set of sliding surfaces as well as by the material composition of the sliding surfaces, and several skills should be brought into the discussions.
b. One or more persons with several years' experience in general problems in friction, lubrication, and/or wear. These people serve as valuable buffers between product engineers, who need a "quick fix," and instrument specialists, who prefer to be more thorough.
c. Specialists in solid mechanics, fluid mechanics, lubricant chemistry, materials science, physics, et al. These specialists must be selected with care, particularly if they are remote from practical problems. Surface scientists, in particular, tend to concentrate on very fine detail, which *seems* sensible, but may not be. Their expertise is vital, however, and can best be applied when problems can be broken down into workable segments by people with broader experience in tribology.

Develop a case history to gain a perspective on how the impressions or convictions were developed that the surfaces in question are operating either properly or improperly. Determine the conditions under which the surfaces seem to behave improperly. If the undesirable phenomenon comes and goes, determine whether this behavior is related, for example, to a change in supplier, a change in weather, a change in the observer, or a change in the process sequence for making the original surface.

Develop a suitable expression for wear rate or performance problems of the surfaces in question. Are the surfaces wearing progressively? Are they scuffing? Is there vibration sometimes but not always? Can these phenomena be quantified?

Decide between examining the wearing surfaces themselves or measuring the *effect* of wearing (or uneven friction, etc.) on the functioning of the machine or component in question. It may be easier or more economical to redesign a machine component to accommodate a particular wear rate or frictional behavior than to find new materials to reduce wear rate or provide more predictable friction. Perhaps both will be necessary. The measurement of component function will probably involve measurement of changes in part clearances, friction, vibration mode, etc. Accommodating a given friction or wear rate is a design question, which will not be discussed further.

If surface examination is necessary, it is useful to plan the steps leading to such examination, as discussed in the next sections.

First Level of Surface Examination

1. Determine, if possible, what effect there will be on the surfaces in question by stopping sliding (eroding, etc), by dismantling the mechanical system containing the surfaces, and by cleaning the surfaces. In some instances the surface chemistry will change with time after the machine is shut off, and surface chemistry will surely change during cleaning. In many instances a test device cannot be stopped, taken apart for examination, and reassembled without making some undesired change.

It is often important to preserve the wear debris on and near sliding surfaces for analysis. Observe the location of build-up of debris, the flow patterns of debris, and the particle size distribution, etc. Obtain oil samples and filter media if the problem is a lubricated system.

2. Dismantle the mechanical system in question in the presence of the person(s) responsible for its performance. Note the practices of the persons doing the dismantling and the possible effect of their practices on surface condition.

3. Use eyes, fingers, and nose to make a first judgment about the environment in which the surfaces are operating. There may be gritty substances or ridges of debris on or near the sliding surfaces, or there may be some particular pattern of marks, pits, or plowed ridges on the surfaces. A 10× eyepiece (magnifying glass) is probably the best aid at this stage.

4. Remove the surfaces to be examined and obtain some wear debris.

5. Observe the surfaces and debris under a binocular microscope that has a magnification range from about 2 to 40. (See Appendix to Chapter 12, Section A.) Use a light source that can be moved to light the target at all angles, from near vertical to near grazing angles. Rotate the specimens under the microscope as well, to observe directional features of the surface.

6. Surface materials may be worn away, rearranged, or built up by transfer. A perspective on these events can often be gained by surface tracing with a Surface Roughness Tracer system (see Appendix to Chapter 12, Section B) or other method of recording surface topography. Weighing of tribological components is useful sometimes. An important point is that the measurement of volume loss (or gain, as by transfer) alone by any of the available methods is not sufficient. Furthermore, measurable weight loss may occur later in the wearing process, so it is better to develop ways of measuring the surface change early in product life. The shape of the worn region, the direction of scratching, the distribution of built-up material, etc. must all be noted.

7. Repeat steps 1 through 6 for several specimens obtained from mechanical systems operated in various ways, with several different materials and with different surface conditions until every observer is sure of the sequence of surface change that is occurring and all agree on the scale (see Appendix to Chapter 12, Section C) of observation needed for full understanding of what is occurring.

Though it is often difficult to do, obtain specimens *in various stages of wear*. When a tribological problem first appears, most investigators become very well acquainted with the failed state of the surfaces. Before the final state, the surfaces have probably gone through several stages of change. Recall that the solution to problems often involves *preventing the first stage of wear or first change in the general behavior of the machine.*

Proceed with patience. Interesting details of the debris and sliding surfaces are usually not obvious in the first hour of study, but with practice, the eye eventually sees differences.

8. Develop a hypothesis on why the surfaces perform the way they do. The best hypotheses will arise from a group of people with the widest knowledge of tribological mechanisms. The hypothesis may contain elements that suggest the need for further analysis of some parts of the system, perhaps by an outside expert. For example, it might be postulated that the problem arises from vibration, or may involve micropitting or hydrogen embrittlement (if the material is hardened steel), or may involve the build up of compacted debris or chemical compounds from a lubricant.

9. With these hypotheses, a choice may now be made between proceeding with laboratory analysis or proceeding with further testing of practical parts. *In most instances, further microscopic or chemical examination will not be as useful as empirically altering some part of the sliding system — the materials, assembly practices, lubricants, etc. — for further testing.*

However, if further examination is necessary, proceed to the next section.

Second Level of Surface Observation — Electron Microscopy

The scanning electron microscope (SEM) is probably the most useful secondary analytical tool for surface analysis in tribology. Most SEMs can cover a range of magnification from 20× to more than 30,000×. One major precaution in the use of the SEM is that an effort must be made to retain perspective of size and scale. Perspective may be lost for two reasons. First, the SEM has a depth of field that is about 300 times larger than that of the optical microscope at high magnification. This has an advantage in that most details on a rough surface will be visible, but a disadvantage in that surfaces appear to be very much smoother than they are. Second, the great temptation when using the SEM is to focus on details that appear interesting, but which often turn out to be irrelevant.

Specimens for the SEM must usually be small, typically no thicker than 20 mm and no larger in diameter than 40 to 100 mm, depending upon the particular brand of SEM. They must be cleaned of volatile substances to an extent depending on the type of SEM (unless the specimen can be cooled to cryogenic temperatures in the SEM). Some operate with a vacuum of better than 10^{-5} Torr (1.33 mPa), but others use pressures nearer atmospheric. If the specimen is a nonconducting material, it must be coated with carbon or gold so that an electron charge does not build up on the surface and deflect incoming electrons.

Images in the SEM do not correspond exactly with what is seen in the optical microscope. The SEM produces an image because the polarities across the specimen surface vary slightly. Regions of + bias appear dark, and regions of – bias (or with accumulated negative charge) appear bright. The optical microscope, by contrast, produces an image of contrasting light reflectivities. It is often useful to compare photos from an optical microscope and an SEM of the same magnification.

SEMs are often equipped with energy dispersive x-ray analysis (EDAX) instrumentation (see Appendix to Chapter 12, Section D.3.e.3) for the purpose of identifying atomic elements in chosen regions on surfaces. Interesting regions may be brought into the field of view in the SEM, small details can be outlined within that field of view, and the elemental composition of the surface material within those outlines can be printed out directly in the most modern and automated instruments.

Sometimes totally unexpected elements will appear in the analysis. This occurs most often when scattered electrons in the specimen chamber impinge on the specimen holder or some other part of the instrument in the vicinity of the specimen, or it may be due to a partially obstructed electron column.

The operation of modern SEMs equipped with EDAX instrumentation does not require high skill. However, a skilled operator should be available to clean, align, and calibrate the instrument on occasion, as well as to aid in the interpretation of some results.

A second type of electron microscope is the transmission electron microscope (TEM). It provides a view *through* a thin layer of solid material of thickness up to 100 nm, depending on the voltage of the electron beam. Specimen preparation requires skill and patience since it is usually done by chemically etching away unwanted material. Surface features of specimens can be observed with the TEM, but this requires making replicas, shadowing, and several other time-consuming steps. High resolution is available in the TEM, but skill is required to do anything with the TEM. Modern TEM instruments are also equipped with electron diffraction instrumentation, which has several advantages over x-ray diffraction.

Selecting Chemical Analysis Instruments

Several steps must be taken to get the results you need.

Decide what information is desired from the surfaces under examination. This is necessary in order to choose the proper type(s) of instruments and to avoid a deluge of costly information. The type of information needed may include the following (with further details given later):

a. Integrity of the original materials due to surface cracks, loose grains, residual stresses, unexpected phases, inclusions, laps and folds from surface processing, etc.
b. Chemistry of "used" surfaces. Oxides, sulfides, organic compounds, decomposed lubricants, foreign matter, mixtures of phases from the original substrate materials, etc., all influence sliding and wearing performance of machine components.

Compare the *capability* of instruments with *information needed* from the instruments. This includes:

a. The scale of depth and width. *No* practical sliding system consisting of material x sliding against or eroded by material y remains in its original state. After a short time each material is coated with other chemical species. If the coating is very thin, e.g., 10 nm, then the analysis of that coating must be done with an instrument that penetrates no deeper than the coating. (See Appendix to Chapter 12, Section D.) On the other end of the scale, if the coating is thick, e.g., 100 nm, and its composition varies throughout its thickness and over its expanse, then one instrument reading taken from the top three atomic layers over a target diameter of 10 atoms will provide data of very limited value.

b. Some analytical instruments identify elements only, and others provide information from the candidate compounds that may be present. Most instruments operate within a limited range of the periodic table of the elements, but are, ironically, unable to identify the most common elements on sliding surfaces, namely, hydrogen, carbon, and to a lesser extent, oxygen. The time required, instrument charges, and operator expertise are usually proportional to the amount of information available from the instruments (as well as time required for the analysis).

c. Several instruments operate with specimens contained in a vacuum. Volatile substances are usually not allowed into these instruments by the operator unless provision is made to cool the materials to a very low temperature. Non-volatile substrates are usually rigorously cleaned before placing them into the vacuum. *Unfortunately, this cleaning removes many of the coatings of interest.*

Develop a statement of precisely what data are to be obtained and how to interpret the results in a manner that is useful to the examination exercise. Instrument operators can explain results in terms of elements and compounds but not always in terms that are useful for solving a failure analysis problem.

Collect and tabulate all information and analyze these data in the light of the hypothesis developed at the outset. From this point the process is obvious and self perpetuating, and will succeed in proportion to the extent of knowledge of wear mechanisms brought into the deliberations.

(See Problem Set question 12.)

Appendix to Chapter 12

INSTRUMENTATION

It seems unnecessary to present this information here because it is available in many books. The major problem is that most books on chemical analysis instruments are written like computer manuals.

A. RESOLVING POWER, MAGNIFICATION, AND DEPTH OF FIELD IN OPTICAL MICROSCOPY

Resolving power is the ability to distinguish two self-luminous points and is calculated to be,

$$R \approx .61\lambda \,/\, (NA)$$

where NA is the numerical aperture of the objective lens of the microscope. (Maximum NA for objectives used with immersion oil is about 1.4, whereas those used in air have a maximum NA of about 0.95.) Numerical aperture is related to the magnifying ability of microscope objectives as seen in the table below.

NA	Magnification (× value)	Working distance	Diameter of field	Depth of field
.2–.3	10	4–8 mm	1–2 mm	10 µm (400 µ in)
.65–.85	40	.2–.6	.25–.5 mm	1–2 µm (40–80 µ in)
1.2–1.3 (oil)	95	.11–.16	.1–.2	0.5 µm

Since few objects are self luminous it is useful to assume that the R will be about twice that calculated above, and taking the value of $\lambda \approx 555$nm (green light) the useful R value for the very best metallurgical microscope operating in air ($NA \approx 0.95$):

$$R \approx 1.2 \times .555/.95 = 0.7\ \mu m$$

This may be compared with the resolving power of the human eye, which is about 50 µm (0.002 inches), limited by the construction of the eye and not by the wavelength of light.

Overall magnification of a microscope is the product of the magnification of the objective times the magnification of the eyepiece. Eyepieces usually have magnifications in the range of 5× to 20×. Generally, there is no point in putting a high-powered eyepiece behind a low-powered objective because it does not aid resolution. You only get a bigger picture.

An important limitation of optical microscopes is the inability to focus upon points of very different elevation in the field of view. This limitation is expressed in terms of depth of field as given in the table above. If the elevation of points on a surface differs by more than the given depth of field, some points will be out of focus.

B. SURFACE ROUGHNESS MEASUREMENT

The measurement of surface roughness has not yet been accomplished to everyone's satisfaction. Most of the disagreement is on the matter of scale — shall we describe a surface objectively according to the arrangement of atoms at the surface, or shall we describe surfaces in terms of how they perform in some situation, e.g., in lubricated sliding or in optical function?

In technology, several methods have been used to describe surfaces, both their locations (part size) and roughnesses. Some of the methods involve the stereo microscope, dark field illumination, reflected light intensity, monochromatic light at lowest angle at which the reflection of the light source appears sharp, capacitance and electrical resistance, electron back scattering, fluid flow, etc. Most systems provide an average roughness over an area, and some will provide profiles in chosen directions. Few of these provide information that is as easy to interpret as data from the stylus tracer.

The world standard method of measuring surface roughness is done with the surface tracer. (This method was invented at the University of Michigan.) Up to the 1980s a tracer system consisted of a spherical-tipped diamond or other hard projection much like a needle or stylus of a record player, with a vertical load applied ranging from 5 mN to much smaller values. The stylus moves along a solid surface, rising over the peaks and descending into the valleys as it moves. The standard radius of the stylus was about 12.7 μm. The rise and fall of the stylus, that is, its vertical motion, is detected by an electronic system and amplified for various purposes.

Since the 1980s some conventional instruments have used styli of smaller radii ($\approx$ 2 μm or less) upon which a smaller load than 5 mN is applied. Still later even more delicate instruments became available which use stylus radii and loads in the "nano" range.

The stylus is (usually, but not necessarily always) moved horizontally along the specimen surface by one of two different systems. The simplest is the sled arrangement, which involves two spheres of about 6 mm radius that ride on (slide over) the same surface as that being measured, but to the sides of the tracer stylus. The sled rises and falls a small amount while sliding over the asperities, mostly following the large-scale waviness of the surface. The second and superior mechanical system guides the stylus on a remote sliding surface of high precision. This system allows good control of traverse speed, usually on the order of 0.1 to 1.5 mm/s, and can measure both waviness and smaller scale surface features, i.e., roughness. The electrical signal is analyzed statistically in various ways, yielding as many as 42 parameters.

Data may be presented in two forms, namely, as roughness parameters, or printed on a strip chart, or both.

1. *Roughness parameters:* After the stylus slides some chosen distance, conventionally about 0.030 inch (0.762 mm), the average height of the stylus tip during the trace is calculated to determine a datum. Then various quantities can be calculated, such as the simple average of the height of the surface features that rise above the datum, referred to as Ra, or the RMS of these heights, referred to as the R_q, and as many as 20 other quantities. Results of measurements of changes in surface roughness during sliding may be expressed in any one or more parameters even though company practice may favor, for example, Ra over R_q. The relationship between any of the available parameters and the function of tribological parts is scarcely known by anyone. Thus the choice of parameter is quite arbitrary, except for communication of data with others.

   Most instruments will resolve roughnesses on the order of 0.025 µm, but specialized ones will do much better. Still more advanced machines can calculate the cross-sectional area of material removed by wearing.
2. *Strip chart recording:* Some instruments are equipped with a strip chart printout that reproduces the measured contour for visualization. Figure 12.1 is a composite sketch of such a trace with several common features of a trace. The resolution of the traces is as good as 0.025 µm as well. One note of caution in using the printout — it is customary to use a much more (100 to 1000 times) magnified vertical scale than horizontal scale on strip chart printouts, which makes the measured surfaces appear to be as rough as the Rocky Mountains. Actually, most surfaces have asperities with slopes less than 10°.

The conventional tracer scratches all but the hardest materials. The mean contact pressure between the tracer tip and a flat *elastic* surface in most older instruments is calculated at about $p_m \approx 6.6E^{0.67}$, where E is Young's Modulus of the specimen surface. Grooves will form in materials with a yield point in tension less than about two thirds of this value, which for steel corresponds to a hardness of about 55 Rc. The grooves will follow the larger hills and valleys of softer material fairly well but will smooth-over the finer surface features.

C. MATTERS OF SCALE

1. Size Scale of Things

When studying objects with microscopes, it is often useful to think about the scale of observation relative to the scale of size of various things. Figure 12.2 is a scale marked in SI and English units.

2. The Lateral Resolution Required to Discern Interesting Features

For observing cracks, defects, inhomogeneities, plastic strains, and the details of surface damage, instruments with an appropriate lateral resolution must be

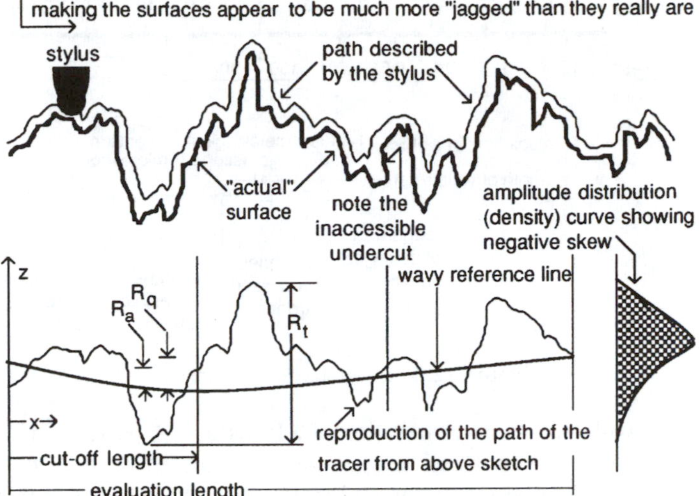

surface tracer instruments sample data over chosen distances in the x-direction, digital instruments divide the distance into smaller units and measure z-values at each interval for statistical operations, producing various numbers, such as

- R_a - the arithmetic roughness average
- R_q - the r.m.s. roughness average
- R_t - the total or peak to valley roughness
- waviness
- amplitude density function
- skewness
- tabulation, if desired, of height of highest peak, number of valleys within some specified evaluation or cut-off length, etc.

Figure 12.1 Sketch of a roughness measurement of a surface.

selected. The question of *appropriate* resolution relates to the "need to know." For example, consider a crack, which nominally we may describe as having the shape of the letter V. At the crack tip the size or spacing may be as small as atomic radii, (≈ 0.3 to 0.5 nm), whereas at the other end it may be visible to the naked eye.

Very narrow cracks may constitute little hazard in a structure and thus may not be worth looking for. However, when a fatigue mode of wear is encountered, even the smallest crack is of interest.

Material defects are of a few atomic dimensions and may be no more useful to observe than are crack tips. Material inhomogeneities are of the order of grain sizes (≈1 μm), and are often more important to find.

The dimensions of wear damage are often large compared with the lateral resolution of instruments. The contact diameter between two hard steel balls of 12 mm diameter pressed together with a load of 450 N is about 0.52 mm. The field of view of a high-power (2000 ×) optical microscope is about 50 μm, whereas the field of view of an SEM at 20,000 is about 5 μm. What will we see? We may

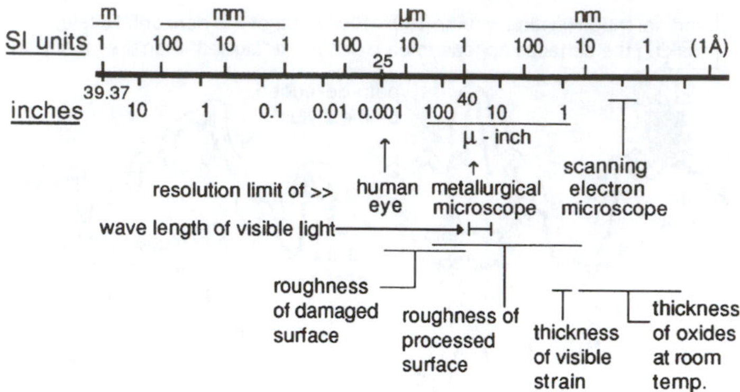

Figure 12.2 SI versus English units.

compare this with the human observation of a landscape. A person may be living within an interesting geological area without realizing it. Assume that such a person is very familiar with a region within 10 miles of his home. If that person could overview a region of 200 miles, he might discern old lake beds and other features, but a higher overview, encompassing an expanse of 1000 miles, may reveal ancient glacier movements. In the same way observations at high magnification are likely to provide more confusion than enlightenment. It is best to get an overview, then focus on special regions. This must be done to represent fairly what is happening. Very often regions of photogenic interest are selected at random providing no good overview.

D. THE CAPABILITY OF CHEMICAL ANALYSIS INSTRUMENTS

1. Introduction

There are more than 100 types of instruments available for identification of chemical elements and compounds in solids, liquids, and gases. They direct either light, x-rays, electrons, ions, neutral particles, or phonons upon a specimen surface. The impinging radiation interacts with the electrons of the target, producing light, x-rays, etc., not necessarily of the same type as the impinging radiation. The emerging radiation is analyzed for its wavelength or energy to determine the composition of the target material. For more information on the capability of analysis instruments, see *Surface Effects in Adhesion, Friction, Wear and Lubrication* by D.H. Buckley, Elsevier, 1981, and *Scanning Electron Microscopy and X-ray Microanalysis*, by J.I. Goldstein, D.E. Newbury, P. Echlin, D.C. Joy, C. Fiori and E. Lifshin, Plenum Press, 1981.

The important questions to ask about any particular chemical analysis instrument are the following:

a. Whether it will identify chemical *elements* only, or chemical *compounds* only, or both.

b. The range of elements on the periodic table that the instrument can identify.
c. The depth of material from which information is gathered and analyzed.
d. Whether the analysis method is destructive or not. Generally, all high energy beams will cause thermal damage to volatile materials. Ion beams cause evaporation of target materials, which may be considered damaging in some instances and a helpful method of removing surface atoms in other instances. The latter is called *ion milling* or *ion beam etching*.
e. Whether the specimens require rigorous cleaning.
f. The size of specimen that can be accommodated by the instrument.
g. The time required from submitting a specimen until the results are available from the instrument.

The choice of instrument is often a matter of which is the most available. A good starting point in choosing a method of chemical analysis is to discuss your needs with the operator(s) of any of the instruments. They may recommend alternatives to the ideal instruments or even alternatives to the instruments in which they have expertise, but much useful information can be obtained from them. Further, they usually know of new variations of conventional analysis instruments. The essential features of the most common instruments are described below to provide a basis for discussion of instruments with specialists. The first six make use of electrons and x-rays, the seventh employs ions, and the eighth uses light. Following is a short discussion of radioactivity.

2. Structure and Behavior of Atoms, Electrons, and X-Rays

a. Basics

An atom consists of a nucleus composed of Z (an integer, also called the atomic number) protons and A–Z (A=atomic weight) neutrons. To balance this nuclear charge, Z electrons are distributed outside the nucleus and revolve around it. The electrons are stationed in specific shells, referred to as the K, L, M, N, etc. shells, counting from the innermost shell outward. There are two electrons in the K shell, up to 8 in the L shell, up to 18 in the M shell, etc. All except the K shell have sub-shells, numbered, for example, L_1, L_2, and L_3 in the L principal shell, etc.

The energy state of an electron in each type of "free" atom (i.e., not combined with another atom) is determined by both the shell it occupies *and* the total number of electrons in the electron configuration of a particular atom. Those nearest the nucleus are in the lowest (negative) energy state. The outermost shell of electrons is in the highest (negative) state of energy, and these are referred to as the valence electrons.

When two atoms are combined to form a compound, the energy state of a valence electron is altered or shifted. If an atom has two or more shells (lithium, Z=3, and heavier), the electrons in the K shell are not significantly affected by bonding to another atom. For hydrogen (Z=1) and helium (Z=2), the K shell is both the inner and outer shell so that when these atoms are combined with others (e.g., hydrogen combined with chlorine to form the acid HCl) the energy state of the electrons in the K shell is affected.

X-rays are photons, as is light, but the more useful x-rays are in the range of λ from 0.5 to 10 nm, whereas the range of λ for visible light is from about 450 to 650 nm. X-rays are generated by bombarding a group of atoms with electrons (as well as other particles). Low-energy electrons displace orbiting electrons partially without removing very many from orbit, which produces a wide and continuous range of x-ray frequencies. These electrons emit x-radiation when they return to stable orbit. Higher energy electrons will completely knock some electrons from various orbits as well as disturbing others in their orbits. Electrons from higher orbits will replace those knocked from a lower orbit, emitting energy in the process. This transition from one discrete shell (and sub-shell) to another emits a very specific quantum of energy in the form of x-radiation, of very specific energy and wavelength. Where several electrons have been knocked from several shells, several frequencies of x-radiation are emitted. These are known as the *characteristic* frequencies or wavelengths of an atom. Each type of atom emits a low level distribution of x-rays but strong emissions at unique frequencies, and atoms may therefore be identified by measuring those frequencies after bombardment by electrons.

b. Obtaining a Stream of Electrons

Electrons in metals migrate freely and some even jump out of the surface. The average distance of jump is proportional to the absolute temperature. This is called *thermionic emission.* When a metal is at ≈2000°C the electrons jump far enough that an electric field of 1000 volts can attract the electrons elsewhere. If a cold target surface is held electrically positive relative to the hot metal, a stream of electrons impinges upon the target surface. The electrons are charged particles and thus electric and magnetic fields can influence the direction of their motion. In analytical instruments (as in cathode ray tubes) the electron stream is focused and directed (and is then called a beam) to chosen locations on a target surface.

c. The Measurements of X-Ray Energy

The energy and frequencies of radiation can be measured by a spectrometer, that is, a device for separating a spectrum into its parts. One type of spectrometer measures energy levels, and these are called energy dispersion spectrometers (EDS). Others measure the frequency of radiation and these are called wavelength dispersion spectrometers (WDS). Each has its advantage over the other. The WDS is slower but more precise than the EDS.

In the EDS, x-rays pass through a silicon crystal doped with lithium for high sensitivity and resistivity. This crystal is cooled to liquid nitrogen temperature to reduce noise due to lithium and to limit electronic noise. Both faces (ends) of the silicon detector are coated with gold and a high voltage bias is applied to the gold. When an x-ray photon strikes the silicon crystal it produces electrons and holes, which are attracted to opposite ends of the crystal because of the bias voltage. The formation of an electron/hole pair requires 3.8 eV, and thus the number of pairs formed indicates the energy of the x-ray. In unit time the collected charge is measured, converted to a voltage pulse, and digitized etc.

The WDS systems place a crystal of known atomic spacing in the path of emitted x-rays. X-rays will be reemitted from the crystal by diffraction at several very specific angles relative to the orientation of the incoming beam. An x-ray counter is located at some fixed position. The crystal is then rotated until radiation activates the x-ray counter, and this is done for several of the strongest wavelengths. The rotational position, θ, of the crystal relative to the impingement direction of the first order x-ray beam is related to the wavelength of the radiation by Bragg's law of diffraction, $\lambda = 2d\sin\theta$ where d is the atomic spacing in the crystal. Since x-rays in the wavelength range 0.1 nm to >1.0 nm may be expected in these instruments (from the heavy to the light elements), there are several appropriate crystals that may be used (from LiF, d = 0.2 nm to potassium acid phthalate, d = 1.33).

The data from each of the EDS and WDS systems are a plot of intensity versus either energy state measured by changing bias voltage or frequency measured as the Bragg angle (or wave number). The plot will consist of peaks, valleys, or steep slopes. The location of these features along the abscissa may be compared with data taken previously from all known elements and compounds and published in large handbooks. Modern instruments use a look-up file in a computer. When a reasonable match has been made, the specimen under analysis is identified. This is a simple exercise where a single element or compound is present. Experience or an expensive computer is required to sort out the peaks (valleys, slopes) of close peaks from mixtures of materials that overlap to form a single new peak.

Data will often be labeled by a code according to the cause of radiation. Electron beams produce K, L, and M ionizations of the atoms and generation of several characteristic X-radiation frequencies. X-rays coming from electrons dropping from the L shell to the K shell are called K_α and those dropping from the M shell to the K shell are called K_β etc. In addition, those dropping from the first sub-shell of the L shell group are called $K_{\alpha 1}$, etc.

d. Electron Impingement

Impingement by electrons, as well as other particles, produces scattered electrons and secondary electrons. Impinging electrons of sufficient energy may eject electrons completely from the shells. Some escaping electrons are made to pass between electrostatically charged plates. The path of the electrons will therefore be curved according to the energy or velocity of the electrons. The voltage of the electrostatic field varies over a given range, progressively directing streams of electrons of different energy into an electron detecting diode. In this manner, a spectrum of discrete electron energy levels can be tabulated. Since every atom type has electrons of unique configuration, the element from which the electrons are ejected may be identified.

3. Description of Some Instruments

a. Instruments That Use Electrons and X-Rays

See the table on page 227 for a comparison of some instrument capabilities.

1. X-ray diffraction determines the crystallographic (atomic) structure of materials. X-ray beams (usually from light elements because they produce narrow lines) about 10 mm in diameter are directed toward a specimen surface at some angle, θ. There are three major methods in x-ray diffraction, varying either λ or θ during the experiment. The methods are:

	λ	θ
Laue method	variable	fixed
Rotating-crystal method	fixed	somewhat variable
Powder method	fixed	variable

Fixed λ values are obtained by diffraction

In each case a great number of cones or dots of radiation is diffracted from the specimen and made to fall upon photographic film, either wrapped around the specimen or placed near it. The angular orientation of these photographically developed spots or streaks around the hole(s) in the negative (through which the impinging beam passes) and their distance from the impinging beam indicates the lattice structure and orientation of the target material.

To analyze thin films an x-ray beam is directed toward the surface at a low angle in order to maximize the distance traveled through the film and minimize the distance traveled through the substrate. The range of impingement angle is necessarily limited, which causes considerable difficulty in resolving the crystallographic structure of inhomogeneous film materials.

2. Electron diffraction may be treated the same way as x-ray diffraction. High energy (>10 keV) electrons penetrate as deeply into material, or will pass through as thick material as do x-rays. Analysis to only a few atoms deep, and of adsorbed substances, can be done with electrons of less than 200 eV energy.
3. Energy dispersive x-ray analysis (EDAX) instruments are often added to scanning electron microscopes. Ordinarily the SEM directs a beam of electrons toward a specimen surface which rasters to cover a much larger area of the specimen surface than the diameter of the electron beam (≈1 nm). In this ordinary mode, the SEM is used to scan a surface in search of areas for chemical analysis. When such an area is found, the SEM can be set into a mode of operation in which the beam focuses on one spot (which can be varied in size). The impinging beam is of sufficient energy (>25 keV) to eject a spectrum of x-rays from the target surface, in accord with the *elemental* composition of the target surface to a depth (≈10 to 100 nm) depending on the incoming beam energy. The energy for each particular wavelength of emitted x-ray indicates the relative amount of particular elements. EDAX is sensitive to as few as 5 monolayers of any particular element and can identify most elements including, and heavier than, boron.

If only known elements are present it is possible to estimate the binding energies of elements, from which first estimates can be made of the compounds present in the target area. This arises from shift in the x-ray wavelength peak for a particular element from that in the pure form. An example of data from EDAX is shown in Figure 12.3.

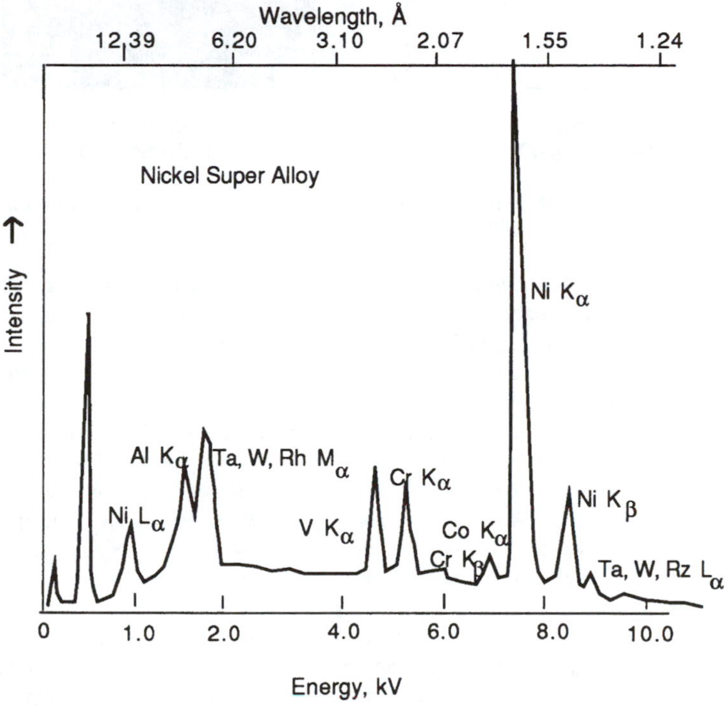

Figure 12.3 EDAX data for a nickel super alloy.

4. Electron micro-probe analysis (EMPA) instruments operate somewhat on the principle of the EDAX in that electrons are directed toward a specimen surface and x-rays from the surface are analyzed. However, the x-rays are analyzed by WDS. The target area is identified by an optical microscope. The impinging electron beam may be set to one fixed location, or be made to raster over a larger area. With the beam focused on one point the diffracting crystal in the WDS is rotated to provide the identity of all elements within its range. With the instrument in the rastering mode the WDS crystal is set to the angle for one operator-determined chemical element. When that element is encountered in the scan the x-ray detector sends a strong signal to a rastering cathode-ray tube from which a photograph of an elemental map is taken. A map of one chosen element is given in one photograph, usually in the form of white spots on a black background. This photograph can be compared with an ordinary SEM photo of the same area to identify various materials in the photo. An example of EMPA scan data is shown in Figure 12.4.

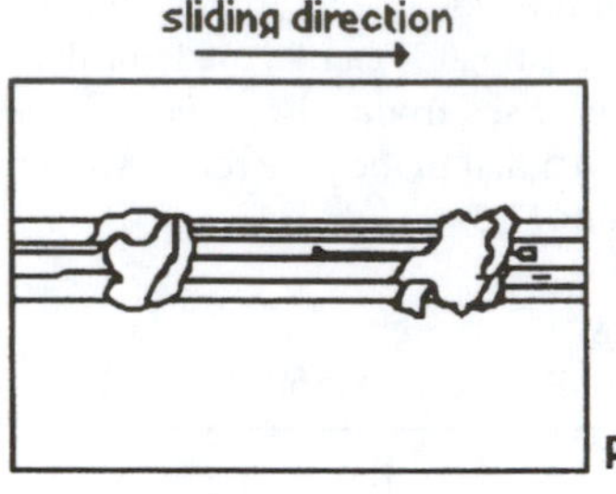

a. Simulated SEM photo of wear track (500x)

b. Simulated map of copper transferred to steel

Figure 12.4 Schematic sketch of a scan of copper (K_α, for example) transferred to a steel surface during sliding.

5. *X-ray photoelectron spectroscopy (XPS)* is also called electron spectroscopy for chemical analysis (ESCA). A typical ESCA instrument uses a specific x-ray source, e.g., Mg K_α (1253.6 eV) at some constant power (e.g., 300 W) in some particular vacuum (e.g., 1.3 μPa or 10^{-8} Torr). The x-radiated specimen emits photo-electrons (same electrons, only now they come out as a result of radiation with photons) from all shells including the core and valence shells. The energies of these electrons are measured with a previously calibrated spectrometer (whose work function is typically set so that, for example, the $Au_{7/2}$ line appears at 83.75 eV and the Cu $2p_{3/2}$ line appears at 932.2 eV). The energy distribution is plotted as peaks. Some peaks will appear at well-known energy levels, indicating the elements present. Other peaks will be near the energy level that corresponds with the valence electrons for elements known to be present. However, the shift from the expected energy level indicates a binding energy when two elements are combined into a compound. This shift has been tabulated for very many compounds. ESCA instruments have excellent resolution of electron energy, and a great amount of effort has been devoted to automating the separating of overlapping peaks of elements. This instrument is among the best available for identifying compounds as well as elements in surface layers of specimens.

ESCA instruments need only 10^{-6} grams of a material and can operate with films that are less than 10 nm thick. It is nondestructive and can identify elements beginning with beryllium and heavier. An ESCA survey profile on a scuffed cam surface is shown in Figure 12.5.

ESCA can be used to provide a depth profile of the composition of thick films as well. Bombardment of a specimen surface with argon ions, using an energy of 3 keV with a current intensity of 8.5 μA and 1 mm beam diameter at a glancing angle of 45°, for example, will remove 1 nm of iron oxide in 10 minutes. An ESCA scan can be done after every 10 minutes or so of this ion milling until a familiar substrate is reached. By this method the thickness of a particular film or layer can be estimated. Scans after the first one will show the presence of argon, which may not interfere with the intended work. Bombardment could also be done with helium, which cannot be identified by ESCA, but it takes 10 times longer than bombardment with argon to achieve the same milling rate.

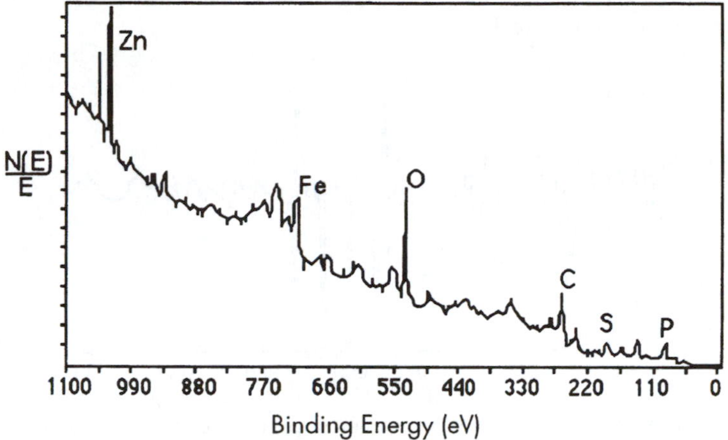

Figure 12.5 ESCA or XPS scan of a film deposited on steel from engine lubricating oil.

After several scans, taken several minutes apart, a plot of the composition profile can be made, as shown in Figure 12.6, for the same sample as Figure 12.5.

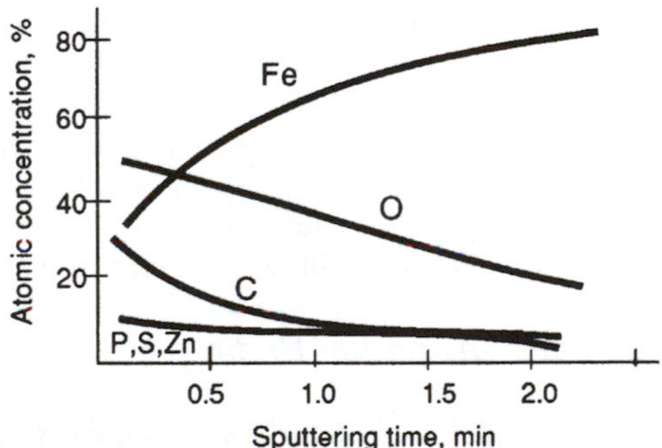

Figure 12.6 Plot of the results of an XPS scan over several minutes of sputtering.

6. Auger (pronounced "OJ") emission spectroscopy (AES) is the ultimate in a *surface* analysis tool. It directs an electron beam in the low range of 1000–3000 eV toward the specimen and measures the energy of the lower energy beam emitted by the specimen. Such low energy electrons penetrate only 4 to 5 atomic layers deep. AES instruments need only 10^{-10} grams of a material. They are nondestructive and can identify elements beginning with lithium. The name of this technique is taken from a particular type of electron emission, in which an incoming electron knocks out an electron in the K shell, which is replaced by an electron from the L shell, which in turn releases sufficient energy to emit an Auger electron from the M shell. The yield of Auger electrons is high for elements of low atomic number where x-ray yield is low. An AES scan of the cylinder wall of a fired gasoline engine is shown in Figure 12.7. Fe, S, C, Ca, and O are identified in various amounts. AES can be used to provide a profile

of the composition of thick films, again by ion milling as described in the previous paragraph on ESCA.

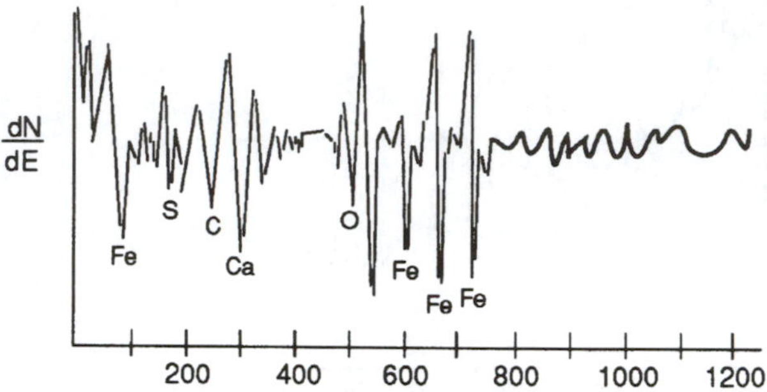

Figure 12.7 Auger scan (differentiated scan) of an engine cylinder wall.

7. Secondary Ion Mass Spectroscopy (SIMS). Particles sputtered during ion bombardment contain information on the composition of the material being bombarded, and the masses of the charged components of these sputtered particles are determined in SIMS using conventional mass spectrometers (magnetic or quadrupole instruments). The sputtered particles (ions and neutrals together) reflect the true chemical composition of a bulk solid even when selective sputtering occurs. Since sputtering largely originates from the top one or two atom layers of a surface, SIMS is a surface analysis instrument. But it is intrinsically destructive. The basic information is the secondary ion mass spectrum of either positive (Na) or negative (Cl) ion fragments. The SIMS spectra typically include peaks for both ion types, but also peaks for the NaCl combined with neutral fragments, appearing as, for example, $NaCl_2$, etc. This method is specifically useful for characterizing different adsorption states.

b. An Instrument That Uses an Ion Beam

Elastic recoil detection is one method that may be used to indicate the presence of hydrogen on the surface of specimens. Helium ions are directed toward a specimen; typically a beam of $\approx$3 mm diameter impinges at an angle of 12.5° from the horizontal orientation. Typically the beam energy is set at 2.4 MeV with, for example, 60 μC beam charge. Both helium and available hydrogen are scattered off the target toward a detector. The forward scattered helium ions are stopped by a film of Mylar and the relative amount of hydrogen is indicated by current in the detector behind the Mylar. Hydrogen concentrations as low as 1% may be detected. The scattered hydrogen ions emerge from only the upper five atomic layers in the film.

c. Instruments That Use Light

Infrared spectroscopy (and the automated version augmented by Fourier transform calculations, FTIR) is most useful in detecting the change in chemistry in liquid lubricants and for identifying organic compounds on lubricated surfaces. In this method infrared radiation, in the range of λ from about 2 to 15 μm (from a heated ceramic material) is directed to pass through a transparent substance (a solid, liquid, or gas) or reflect from a highly reflecting surface on which there is a transparent substance to be identified. This method requires little specimen preparation and is usually operated in air, except that in the reflected mode the solid surfaces must be fairly smooth.

The spectrum of the radiation that passes through or is reflected from the specimen is recorded and compared with that coming directly from the source. In some instances, if some opaque liquid is diluted in a solvent the radiation through the combined liquids is compared with that which passes through the solvent only.

In the simpler instruments a plot is provided of the % absorbence (alternative presentation of data, % transmission) of radiation versus frequency (cm^{-1}), or perhaps a ratio of % absorbence of solvent with and without sample liquid. A spectrum for solid polyethylene is shown in Figure 12.8.

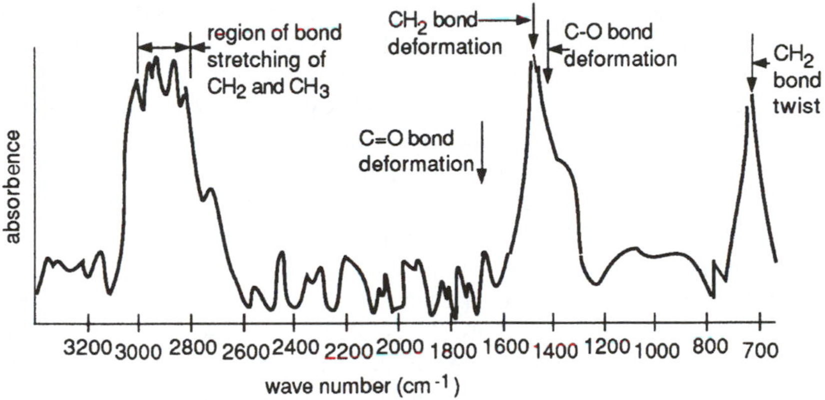

Figure 12.8 Infrared scan of solid polyethylene, with major peaks identified. The location of peaks for C–O and C=0 are shown, and the scan indicates that these bonds are not prominently represented in polyethylene.

In modern instruments, the data are analyzed by Fourier transformation in order to determine whether an irregular curve may be the sum of two overlapping absorption bands. The data ultimately indicate the existence of the linear, rotational, and coupled vibration modes of bonds between atoms. Light will be absorbed when its energy is transformed into vibration of those bonds. Since every compound is made up of arrays of bonded atoms the infrared absorption spectrum becomes a tabulation of the relative number and type of atoms and bonds in the specimen.

Computerized instruments will read out all possible compounds that may be contained in the specimen. The resolution is adequate to identify monolayers of CO on metals or parts per thousand of long-chain hydrocarbons in solvents, for example.

The presentation of data in terms of frequency, cm^{-1} is not readily understood. Actually, it is simply $1/\lambda$, (omitting the velocity of light, c) having the proper units of reciprocal-length but referred to as frequency. A complete spectrum may cover the range between 4000 cm^{-1} and 2 cm^{-1}, but common instruments cover up to 400 cm^{-1}. Often a plot is presented covering a narrower range, for example, 2200 cm^{-1} to 1700 cm^{-1} if most of the difference between two specimens appear in this range.

A comparison of instruments can be found in Table 12.1.

4. Ellipsometry and Its Use in Measuring Film Thickness

Effective breaking-in of lubricated steel surfaces has been found to be due primarily to the rate of growth of protective film of oxide and compounds derived from the lubricant. The protection afforded by the films is strongly dependent on lubricant chemistry, steel composition, original surface roughness, and the load/speed sequence or history in the early stages of sliding. Given the great number of variables involved, it is not possible to follow more than a few of the chemical changes on surfaces using the electron microscopes and other analysis instruments at the end of the experiments. A method was needed to monitor surfaces during experiments and in air. Ellipsometry was used for real-time monitoring, and the detailed analysis was done by electron-, ion-, and x-ray-based instruments at various points to calibrate the results from the ellipsometer.

A complete description of ellipsometry may be found in various books and the particular ellipsometer used in the work mentioned in this article is described in reference 6 of Chapter 9. Fundamentally, ellipsometry makes use of various states of polarized light. The effect of a solid upon changing the state of polarized light is now described.

Polarized light is most conveniently described in terms of the wave nature of light. Plane polarized light is simply represented as a sine wave on a flat surface as shown in Figure 12.9. The end view of the wave may be sketched as a pair of arrows. Light is directed toward a surface at some chosen angle relative to a reflecting surface, called the angle of incidence as shown in Figure 12.10. If the surface is no rougher than about λ/10, the light will reflect with little scatter, which is referred to as specular reflection. Linearly polarized light may be directed upon a surface at any angle of rotation or azimuth, relative to the plane of incidence as shown in Figure 12.11. The incident light can be represented as having separated into two components, the s component and the p component.

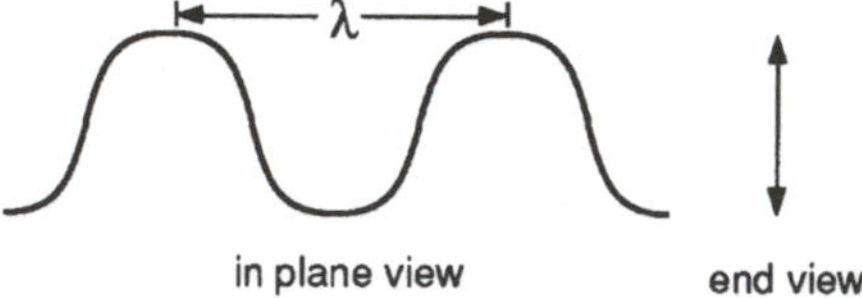

Figure 12.9 Sketch of a wave of radiation, in plane view and end view.

Each component is treated differently by the reflecting surface: each component changes in both phase and intensity at the point of reflection, as shown in Figure 12.12. The sketch shows an incident plane polarized beam at an azimuth of 45°,

Table 12.1 Comparison of Four Main Chemical Analytical Instruments (all require a vacuum) (z of 1=H, 2=He, 3=Li, 4=Be, 5=B)

	EMPA WDS	EDAX EDS	AES	XPS (ESCA)	SIMS
Incident particle	Electrons 10–30 KeV	Electrons 10–30 KeV	Electrons 1–20 KeV	X-rays 1.25–1.49 KeV	Ions 0.5-20KeV
Emitted particle	X-rays 2.5–15 KeV	X-rays 2.5–15 KeV	Auger electrons 1–350 KeV (energy of emitted electrons)	Photoelectrons 20–2000 KeV (energy of emitted core level and Auger electrons)	Secondary ions
Element range	>B (z=5) (quant)	qual.(4<z<11) quant.>Na (z=11)	>Li (z=3)	>Li (z=3)	H-V
Elemental resolution	0.1%	1%	0.1%	0.1%	0.01%
Depth of analysis	1μm	1μm	2nm	10nm	2nm
Lateral resolution	1μm	1μm	>20nm	>150nm	>100 nm
Information provided	These provide good information on elemental composition with some indication of chemical structure			Elemental + chemical structure	Best elemental specificity
Elements not identified	H, He, Li, Be, B + C,N, O, F, Ne		H, He	H, He	None
Surface coating required?	(For nonconducting materials)			No	No
Pump down time	15 min.	15 min.	10 hours	30 min.	30 min.
Measurement time	1 hour	100 sec.	30 min.	30 min.	2 hours

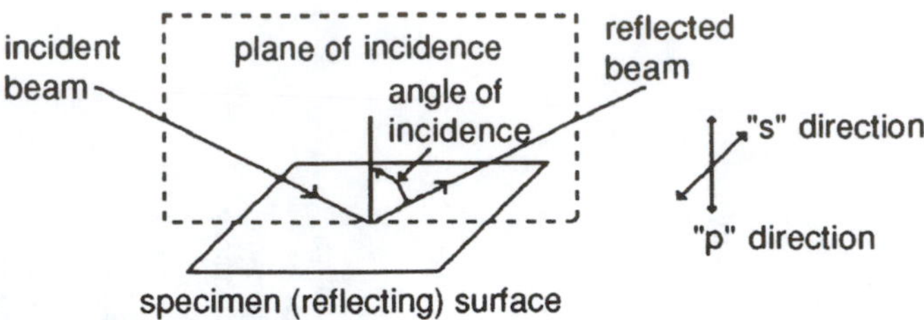

Figure 12.10 Sketch of a light beam incident upon and reflecting from a surface, showing the plane of incidence and the directions of the "s" and "p" components of light.

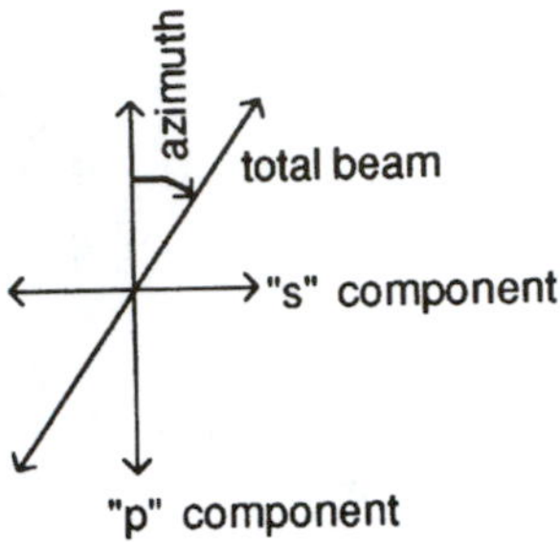

Figure 12.11 Sketch of one possible orientation of a plane or linearly polarized beam relative to the plane of incidence (which contains the "p" component).

which can be thought of as separating into two equal components. Each component is changed in both intensity and phase at the point of reflection. The reflected components then recombine to form a beam that is polarized elliptically. The elliptical shape may be seen by plotting the s and p components, both at the same time or point in the reflected waves over a wavelength. (The incident and reflected beams are shown in line for convenience in visualizing the different phase shifts of the two beams.) Formally, the changes in intensity of each of the s and p beams, and the phase shift, δ, of each are expressed as follows. The intensities the incident s and p waves may be expressed as E_s and E_p, and the intensities of the corresponding reflected waves as R_s and R_p: the absolute phase position, δ, of each of the incident and reflected s and p waves may be expressed with the proper subscripts, the ratios are defined as,

$$\frac{\dfrac{R_p}{R_s}}{\dfrac{E_p}{E_s}} \equiv \tan\psi \quad \text{and} \quad [(\delta_p)_r - (\delta_s)_r] - [(\delta_p)_i - (\delta_s)_i] \equiv \Delta$$

then $\rho = \tan \Psi\, e^{(i\Delta)}$. The last equation is the fundamental equation of ellipsometry.

Figure 12.12 describes the geometric manner of conversion from one form of polarized light (linear) to another (elliptical). In general the incident beam is also elliptically polarized. Actually, linearly polarized light is simply a special case of elliptically polarized light, as is circularly polarized light. Ellipsometry

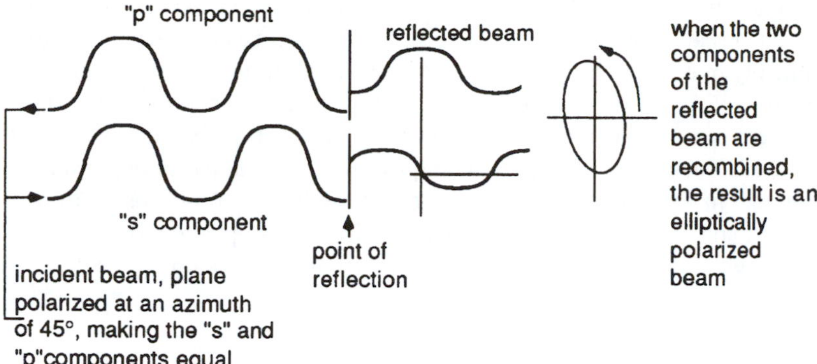
"p" component
reflected beam
when the two components of the reflected beam are recombined, the result is an elliptically polarized beam
"s" component
point of reflection
incident beam, plane polarized at an azimuth of 45°, making the "s" and "p"components equal

Figure 12.12 Sketch of the two separate influences of a reflecting "s" face on the "s" and "p" components of incident plane polarized light.

is the technique of measuring changes in the state of polarized light and using these data to determine the complex index of refraction of specimen surfaces. The complex index is composed of two components, real, n, and the imaginary, κ. The latter, κ, is related to the absorption coefficient.

Ellipsometers can be used to measure either the complex index of refraction or the thickness of thin films on substrates. In the latter case, if the film is thin enough for light of significant intensity to reach the substrate, the film alters the apparent complex index of the system. The influence of the film will be to alter the apparent index from that of the substrate in proportion to the thickness of the film. A measurement taken with light of one wavelength at a single angle of incidence requires knowledge of the complex index of refraction of both the film material and the substrate material. If the index of refraction of either the film material or the substrate is unknown, measurements must be made with either two different colors of light or at two different angles of incidence. If the index of refraction of neither the film material nor the substrate is known then measurements must be made with three colors of light or at three angles of incidence, or some combination. Further, if the film consists of two layers, then even more colors of light or angles of incidence must be used. The colors of light and the angles of incidence must be selected with great care for adequate resolving ability of ellipsometry.

5. Radioactive Methods

Some isotopes (variants) of atoms spontaneously emit energetic particles. This phenomenon is called radioactivity. The particles are of three types, α, β, and γ. The α particles (rays) are the same as the nuclei of helium in that they have a mass four times and a positive charge twice that of a proton. Their velocities range from 1 to 2×10^7 m/s. The β particles are electrons with a velocity approaching that of light, and they have a penetrating power 100 times that of the α particles. The γ particles are neutral and of the nature of short wavelength x-rays. They are the most energetic and harmful, with a penetrating power 10 to 100 times that of the β particles.

Radioactivity is a nuclear property and does not involve the valence electrons. Thus an unstable (radioactive) isotope of an element (Fe_{56}, for example) acts chemically like its stable equivalent (Fe_{52}). Thus a familiar salt or compound can be made with the isotope, then as it wears or transfers to a mating surface its movement can be traced by a detector of one or other of the emitted rays. This is called radio tracing. Another technique, autoradiography, is done much like "x-raying" in medicine. A radioactive source is located on one side of a solid specimen and a photographic film is located on the opposite side. In this case the photographic film is selected to respond to the β radiation rather than to the γ (x-ray) radiation.

CHAPTER 13

Coatings and Surface Processes

Coatings add desirable properties to surfaces, but also detrimental properties that offset some of the desirable ones. Some compensation can be made for the offset, but additional considerations are the strength of adhesion of the coating to the substrate and the effect of the size of the tribologically applied stress region relative to coating thickness.

INTRODUCTION

There are relatively few products on the market that are single components and made of homogeneous materials. Examples include nails, cups made of foamed styrene, concrete blocks, steel beams, and rope. It is instructive to visit a shopping center to see how few such products there are.

The great majority of products are assemblies of two or more obvious and separable components, each selected to fulfill some of the desired attributes of the assembly. For example, a durable shoe is, in essence, a composite structure consisting of a wear-resisting sole attached to a flexible upper segment. The versatility of such products is limited only by the designer's imagination and knowledge of materials and ways to attach the separable parts together. The availability of such products is limited by economics, however, mostly by the high cost of joining materials together. Thus there have always been efforts to achieve desirable properties in single components by making the surface different from the substrate. The substrate is usually expected to provide mechanical strength, ductility, conductivity, and several other functions. The surface is expected to perform very different functions, namely, to resist wear and corrosion, and to have an acceptable appearance, among other things. This chapter discusses surface processing, where the intent is to achieve properties different from those provided by the substrate. This chapter does not include methods of surface finishing for achieving texture or topography, but it does include such surface finishing processes as painting.

Surface processes can be broadly classified in terms of *surface treatment, surface modification,* and *surface coating*. Short examples in each of these groups are listed below, with longer discussions following:

1. Surface treatments are the processes by which surface properties are changed separately from those of the substrate. Perhaps the most common example is found in steel. A piece of 10100 steel can be annealed throughout to achieve a hardness of 250 VPN (Vickers Pyramid Number). The surface can then be heated to 730° C by a flame or a laser to some shallow depth and cooled quickly to produce martensite of 800 VPN hardness. There is no change in chemistry, only a difference in hardness due to heat treatment.
2. Surface modification processes are those that change the chemistry of the surface to some shallow depth, ranging from fractions of a μm to about 3 mm. One old method adds carbon to the austenitic form of low carbon steel, by diffusion. When the entire part is cooled quickly (quenched in water), the substrate remains tough because of its low carbon content, and the surface becomes hard because of its high carbon content. A newer method implants nitrogen and other ions into metals with the effect of distorting the lattice structure near the surface, thereby hardening it.
3. Surface coating processes build up the dimension of some region of a surface. All types of metals, polymers, and ceramics are used as coatings, and they are applied to all types of substrates. Surface processes are many and varied, and are applicable to virtually all materials. Data on prices and properties for purposes of evaluating these processes cannot be put into a convenient table; available information for specific production problems should be obtained from vendors of the machinery and suppliers of such processes. Unfortunately, surface processes are often advertised in the same manner as is laundry soap, including testimonials from shop foremen and sundry purchasing agents. An interested process engineer should assess processes by testing them on actual production materials. Before such tests, however, it is well to become aware of the fundamental events that take place in each process. These are described in the next sections.

SURFACE TREATMENTS

Virtually all processes that change bulk properties will also change only the surface properties, if properly applied. The properties of some materials are changed by heat treatment; the properties of others may best be changed by plastic flow. A partial list of surface treatments is given in two groups, namely, those that use heat and those that plastically deform.

Heat treatment is affected by heating at any convenient rate, but by cooling at controlled rates. The major heat sources are listed below in order of potential increasing surface heating rate. The higher the rate of heating, the thinner will be the heated layer, where the goal is to reach some specific surface temperature. A thick layer will resist wear and indentation longer than will a thin layer, but a thin layer will produce less part distortion than does a thick layer. Note that processes are often given names that only partially describe what takes place.

For example, laser hardening of steel implies that a laser hardens steel. In fact, the laser only heats the steel, after which fast cooling (either in water, or by conduction into the substrate after the heat source is removed) causes the hardening.

a. Flame hardening uses a gas-fired flame, usually oxygen-acetylene, propane, or other high-temperature fuel. This process can be quickly installed, but it is not as readily automated as some others, and it cannot be focused upon very small regions on a surface. The intensity of radiation impinging on a surface from an oxyacetylene torch is on the order of 10^6 to 10^7 W/m^2.
b. Induction hardening is done by placing a metal into a loosely fitting coil, which is cooled by water and in which an alternating high current (from 60 Hz up to radio frequency, i.e., kHz) flows. The current in the coil induces a magnetic field in the metal, which because of magnetic reluctance causes heating in the metal, mostly in the surface at the higher frequencies. The coil current is shut off and cooling water is applied to the part at the appropriate time. This process is clean and readily automated, but it is restricted in its ability to heat specific regions on a surface.
c. Heating in some instances can be done by radiation from an electric arc as in arc welding. The intensity of radiation impinging on a surface from an arc is on the order of 10^7 to 10^8 W/m^2.
d. Heating of surfaces can be done by directing a plasma toward a surface. The plasma is a stream of ions which revert to molecular gases upon approaching a cold surface. The intensity of radiation impinging on a surface upon which a plasma is directed is on the order of 10^9 to 10^{10} W/m^2.
e. Laser hardening uses a laser for heating a surface. The usual wavelength is in the infrared, in the range longer than 1000 μm or 1 mm. The CO_2 laser ($\lambda \approx$ 10 mm) is commonly used. It directs heat of intensity on the order of 10^{10} to 10^{11} W/m^2 upon a surface. A laser system is expensive to install (and is very inefficient in terms of energy taken from the source and applied to heating the target surface, $\approx$ 6%), but the beam is easily steered or directed along any path on a surface by automatic control of mirrors, even into regions that are out of direct line of sight.
f. Electron beam hardening uses a stream of electrons to heat a surface. Industrial safety considerations limit the electron accelerating voltage to less than 25 kV to prevent high emission of x-rays. It supplies a beam of intensity on the order of 10^{11} to 10^{12} W/m^2. The beam can be steered by a magnetic lens but only in line of sight. Conventional electron beam systems require that the part being processed should be placed into and removed from a vacuum chamber ($\approx$1 to 10 mPa). This usually requires some time and skill to operate and obviates the use of fluids to cool a heated part. At higher cost, one can purchase an electron beam system which directs a beam from the vacuum enclosure through an orifice into the atmosphere, for a short distance, where part handling and cooling can be done conveniently. This beam cannot be steered through large angles, and thus the part must be moved about under the beam. Where cooling of a surface is required, after heating, in order to cause a phase change, it may be necessary to do so by quenching in liquid or by spraying liquid on the hot surface. However, a very thin layer of heated material will also cool quickly by conduction to the substrate, if the temperature gradient and the thermal

conductivity are high enough. For example, the conduction cooling that follows heating by a laser or by the electron beam can be sufficient to produce martensite in 1040 steel, but this will not occur when the surface is heated by a flame.

Some plastic flow processes include the following:

a. Burnishing involves pressing and sliding a hardened sphere or (usually) roller against the surface to be hardened. It is a rather crude process which can leave a severely damaged surface. Lubrication reduces the damage.
b. Peening is done either with a heavy tool that strikes and plastically indents a surface, usually repeatedly, or by small particles that are flung against a surface with sufficient momentum to plastically dent the surface. The latter is called shot peening if the particles are metal of the size of ballistic shot. The velocity of shot or other particles may be as high as 35 m/s. It is, therefore, a very noisy and dangerous process.
c. Skin pass rolling is done with spheres or (usually) rollers of a diameter and loading such that the surface to be hardened is plastically indented to a small depth. Large rolls will plastically deform thin plate or sheet throughout the thickness, but skin pass rolling can be controlled to plastically deform to shallow depths.

The local plastic flow that occurs in these processes expands an element of material laterally and thins it, with the effect of developing a compressive residual stress in the surface. A bar that has been shot peened, for example, will bow so that the peened surface will be on the outer radius.

The hardness of a surface that has been severely plastically deformed can be calculated from the tensile stress–strain properties of the material. The more ductile metals can be hardened the most.

(See Problem Set question 13.)

SURFACE MODIFICATION PROCESSES

Surface modification processes are those that change the chemistry of existing materials in the surface of the original material. These include the following:

1. Carburizing is done to increase the carbon content of steel. The maximum hardness of a piece of steel is related to the carbon content. For structural purposes a steel of less than 0.4 percent carbon is desired for toughness, but for wear resistance and indentation resistance a carbon content of about 1 percent is desired. The carbon content of steel can be increased only when the steel is in the austenitic or face-centered cubic state where the maximum solubility of carbon is about 2 percent (at 1130°C). Thus when steel is heated in an atmosphere rich in carbon, some of the carbon will diffuse into the steel. A carbonaceous atmosphere is achieved by using CO, by burning fuel gas with inadequate O_2, or by heating chips of gray cast iron, which usually contains over 2.5 percent carbon. A very rich carbonaceous atmosphere will usually

produce a steep gradient of carbon content in the heated part, which results in large stress gradients and possible cracking during heat treatments. A lean atmosphere adds carbon slowly. The proper depth and thickness of the carburized layer is controlled by temperature and atmosphere. However, precautions must always be taken to prevent oxidation, (atomic) hydrogen diffusion, grain growth of the steel, and undesirable migration of alloying elements in the steel.

Carburized layers of any thickness can be obtained, but the usual thickness is in the range of 1 to 3 mm.

2. Carbonitriding may be done either in a gas atmosphere of ammonia diluted with other gas, or it may be done by inserting a piece of steel into a salt bath, which is a molten cyanide salt or compound. The cyanide supplies both carbon and nitrogen for diffusion into iron, which itself must be in the austenitic state. The role of the carbon is described above. The nitrogen that diffuses into the steel forms nitrides — iron nitrides, but also nitrides of such alloys as aluminum, chromium, molybdenum, vanadium, and nickel — producing a hardness between 900 and 1000 VPN.
3. Ion implantation is done in a vacuum on the order of 10 μPa. Many types of ions may be inserted into a wide range of surface materials in this process, but the easiest to describe is nitrogen in iron. Nitrogen gas is ionized in an electric field gradient of 10^5 volts/mm. The ions are accelerated to a high velocity in a field on the order of 100 KeV toward an iron surface for example, held electrically negative. The usual area rate of impingement of ions is on the order of $10^{15}/mm^2$. As ions enter the iron surface, several iron atoms are evaporated from the surface, and a channel of atoms is displaced to accommodate the stream of nitrogen ions. The nitrogen concentration builds up to about 15 to 20 atomic percent with a peak concentration at a depth of about 0.7 μm for the given conditions. An implanted surface is in a compressive state of stress, which will usually increase the fatigue life of the surface. The surface is also harder but very thin. Implantation affects the corrosion properties of metals in complicated ways and increases wear resistance for some forms of mild wear.

COATING PROCESSES

A very significant coating industry has developed which offers as many as 60 coating processes. Most of the processes can be broadly classified as given below. No attempt is made to name the processes, because in most cases the process is named after the machine that applies the coating or is given the name of the inventor. In the following paragraphs several processes will be described in terms that will lead to an understanding of the vital information an engineer needs concerning a process, namely, the quality of the product. Information on cost must be obtained from the suppliers of coating services. There are very many suppliers, ranging from substantial industries to part-time, home-based operations. The broad categories of processes include the following major ones:

Weld overlaying is done with all of the heat sources mentioned above, but most often by arc and by gas flame. Welding produces very strongly adhering layers, which may be built up to any desired thickness. For corrosion resistance the filler or coating material may be a stainless steel, and for wear resistance the filler may incorporate nitrides and carbides. Soil-engaging plow points and mining equipment are often coated with steel filler materials containing particles of two forms of tungsten carbide, WC and W_2C, which have a hardness on the order of 1800 VPN.

Spraying of molten and semi-molten metals and ceramics is done in air or in low vacuum. The durability of the product depends primarily on the strength of the bond between the coating and the substrate, which in turn depends on how much of the absorbed gases, oxides, and contaminants found on all commercial surfaces are removed or displaced so that the sprayed material can bond to the substrate of the target material. *Oxides are not displaced or removed by these processes*, which constitutes a significant limitation of bond strength. Several processes are described:

Molten metal, usually aluminum, is sprayed in order to coat steel pipe and tanks exposed to weather and to coat engine exhaust systems. The metal doubtless begins to travel from the "gun" to the target in the molten state, but some of the droplets cool to the two-phase region of the equilibrium diagram before they reach the target. This transition is not instantaneous because a phase change entails the evolution of some heat. In any case, the spray travels at various speeds, usually less than 30 m/s. If the spray were solid, the particles would bounce off the target. Liquid would wet a solid surface and solidify, but two-phase droplets partially flatten against the target surface and remain attached partly by wetting forces due to the liquid phase of the spray. A wet snowball hurled against a wall behaves the same way. Upon solidification some other bonding mechanisms must be involved, however. Recall that all solid surfaces are covered with adsorbed gases. The hot sprayed metal, upon striking the target surface, will cause desorption of some of the water. A bond is therefore effected between the sprayed metal and the oxide on the metal substrate. Later the sprayed metal contracts and produces high residual stresses at the bond interfaces, which will limit the adhesive strength of the film to the substrate. But practically, sprayed coatings are fairly durable against very mild abrasion. Their effectiveness against corrosion depends on their continuity. Here again, one can pile drop upon drop from the spray, but the drops must fit tightly together to prevent the incursion of acids and other corrosive substances. Each drop will bond to another through an oxide film, and there will be high residual stresses because of differential contraction from one drop to another.

The coating of surfaces for wear resistance is a fast-growing industry. One process uses spray which is produced by feeding a powder into the flame of a gas-fired torch or through a plasma. The powder can be a mixture of dozens of available metals, ceramics, and intermetallic compounds, selected both for cost and wear resistance. The spray velocity is in the range of 150 to 500 m/s, and the adhesion strength of the sprayed material reaches the order of 70 MPa, which is adequate for many tasks but not for severe abrasion. One process achieves a velocity as high as 1300 m/s of particle impingement, by detonation of a fuel gas in a tube containing a powder of the coating material. The high-velocity particles from such a device apparently remove a large amount of adsorbed water and other contaminants but not oxides. Perhaps there is also an effective packing of particles in the layers of coating. This type of coating appears to have a strength of attachment in excess of 140 MPa, which makes it much more suitable than other processes for abrasion and erosion resistance.

Paints and polymers are in a class of coatings usually applied for appearance and for mild corrosion protection but not for significant wear resistance. These materials are applied to a surface by the spraying, wiping, or rolling of liquid. For effective bonding the surface to be coated must be clean and the liquid coating must wet the solid surface. The coating is then expected to solidify, either by the evaporation of a solvent or thinner from the coating, or by other mechanisms of polymerization of the molecules.

Surfaces can be coated by electroplating, usually in the range from 0.5 μm to about 0.25 mm thick. The common coatings are chromium, nickel, copper, zinc, cadmium, tin, and molybdenum. Some coatings are hard and provide wear resistance. Some are soft and provide protection against scuffing, while others are well suited to protection against corrosion. The process is done in an acid to be plated (for example, a nitrate, a sulfate, or others). A few volts are applied with the part to be plated as the cathode (–). The plating ion concentration, the bath temperature, and the applied voltage must be carefully controlled to avoid poor adhesion of plating to the substrate, spongy plating, or large crystals in the plating. Overvoltage must be avoided because it produces hydrogen, which embrittles some metal. In addition, since the plating thickness is proportional to the current density, some care must be taken in part design, anode geometry, and shielding to make the plating of the proper thickness in all areas. The strength of attachment is high because oxides are removed before metal ions approach the substrate.

Electroless plating is a process that was developed to overcome some of the difficulties of electroplating. (One major difficulty with electroplating is the disposal of the acids used in the processes.) Coatings of nickel–phosphorus or nickel–boron alloys may be applied to a wide range of metals and alloys. Plating occurs by hydrogenation of a solution of nickel hypophosphite, usually available commercially with proprietary buffers and reducing agents. Coatings of any thickness can be applied. The applied coating has a hardness of ≈500 VPN, and the hardness increases to ≈900 VPN when heated to 400°C for one hour. Again, the bond strength is high, if oxide has been adequately removed.

Impregnated coatings are not strictly coatings but are usually classified as such. They are formed by direct contact of the surface to be coated with a solid, liquid, or gas of the desired element (and diffusion occurs through oxides, et al.). An alloy forms in the surface of the part to be coated, which has different properties than that of the substrate. The catalog of such processes is large, including calorizing (Al), carburizing (C), chromizing (Cr), siliconizing (Si), stannizing (Sn), and sherardizing (Zn).

Another process that is not strictly a coating involves the melting of a thin layer of a metal part (in a controlled atmosphere) and then sprinkling TiC or other hard compounds into the molten layer. Upon solidification the TiC becomes firmly bonded and increases wear resistance.

Physical vapor deposition (PVD) is a process that is done in a vacuum of about 10 mPa. The coating material is heated and evaporated (boiled). This vapor fills the enclosure and condenses on cooled surfaces, including the part to be coated. Coatings of any thickness up to about 100 μm may be applied. The adhesion to the surface (often called the substrate) depends on the cleanliness of the surface, but PVD coatings are readily rubbed off unless the coated part has been heated for some time, allowing diffusion of some of the coating into the oxide on the part surface. Ion beam bombardment can be used before deposition to remove oxide and during deposition to form desirable atomic and molecular structures in the coating. TiN is one coating of several that are applied in the PVD process. The vacuum enclosure contains resident nitrogen plus a few percent of argon, krypton, or other gases. Titanium is boiled off, combines with the nitrogen, and condenses. Ion bombardment adds sufficient energy to heat the substrate and activates the Ti and N atoms to fall into the desired lattice structure. TiN may be formed into several lattices, each with its own color. The coating is usually dendritic in structure as well, particularly if the process has proceeded at a high rate. TiN can also be deposited by the CVD process described below, but this requires heating to the point of tempering martensite, thereby causing part softening and probably distortion.

Chemical vapor deposition (CVD) takes place in a "vacuum" of about 10 to 100 mPa. The enclosure also contains a gas, which includes ions of the type to be deposited on the part surface. There usually is sufficient chemical reaction to remove oxide and bond the coating to the part. Chemical reaction occurs at the surface of the base metal M′, with deposition of the coating metal M. There are three types of reactions:

1. When the coating medium or vapor is a chloride (for example),

$$MCl_2 + M' >< M + M'Cl_2$$

2. By catalytic reduction of the chloride at the base metal surface when the treating atmosphere contains hydrogen

$$MCl_2 + H_2 <> M + 2HCl$$

3. By thermal decomposition of the chloride vapor at the base metal

$$MCl_2 <> M + Cl_2$$

The last reaction appears the simplest,but thermodynamically it is often not possible nor very economical. Specialists in these processes should be consulted on such details.

An intriguing development in the late 1980s was the deposition of diamond coatings by CVD. The low pressure atmosphere contains H_2 and CH_4 (methane) mostly. Pure diamond, the tetrahedral crystalline atomic structure with sp^3

bonding, requires a lengthy cycle of forming very thin films, followed by heating to evolve hydrogen from those films. The lower-cost, noncrystalline diamond with high hydrogen content is somewhat less hard than crystalline diamond and adequately satisfies most needs.

Diamond is attractive as a wear-resisting coating because of its hardness, but the problem with diamond coating at this time is that each grain (crystal) of diamond grows independently of the others, which finally produces a coating that looks like abrasive paper. Developments will continue however, not because of a potential market for wear-resisting coatings, but rather because diamond has a very high thermal conductivity and is attractive as a substrate for diodes.

QUALITY ASSESSMENT OF COATINGS

Process variabilities strongly influence the quality of coatings, particularly those involving complicated chemical dynamics. Thus several simple tests are used to assess quality. Coated metal strips are bent or stretched until the coating cracks or flakes off. Hardness tests are adjusted to apply increasing loads until coatings crack. Various shapes of styli are dragged over coatings and the resulting damage is viewed.

One prominent scratch tester applies an increasing load upon a stylus as a coated specimen moves under the stylus. A microphone is attached to the stylus, and when it detects a high level of vibrations the coating is presumed to have failed. The load at which this occurs is taken to be a "figure of merit" of the bond between the coating and the substrate. As with most other tests, this one is useful in production control, but can only remotely indicate the wear properties of the coating. The load at which coating becomes detached, cracked, or flaked off is dependent on the friction between the coating and stylus, the ductility of the substrate, the thickness of the coating (relative to the radius of the stylus), and doubtless several other variables.

The wear resistance offered by a coating should be measured under conditions near those of practical systems. One condition not usually measured by the scratch tester is repeat-pass sliding. Data for coatings of TiN on hard steel show that for five passes of a stylus, the load that causes significant cracking of the coating is about 10–15% that for a single pass. Microscopic cracks appear in the first pass, then propagate, link with interface cracks, and finally lead to loss of the coating.

CHAPTER 14

Bearings and Materials

INTRODUCTION

Ball and roller bearings are referred to as rolling element bearings (formerly known as antifriction bearings). Their usual competitor is the sliding bearing, the simplest of which is a shaft turning in a sleeve or drilled hole. Rolling element bearings are very prominent in our technology, somewhat out of proportion to their real advantage over the plain bearing. It was probably first the bicycle and then the automobile that provided the main driving force for the development of the rolling element bearing industry. The automobile evolved from the wagon and buggy, which had sliding bearings in the wheels, but these bearings were not reliable at higher speeds or with only minimal maintenance. Thus rolling element bearings were introduced, and today all automobiles contain some. Whereas the automobile propelled the development of rolling element bearings, it is in the automobile that their proper economic place is seen. Most bearings in engines and transmissions are sliding bearings. Likewise sliding bearings are prominent in high volume items such as low-cost electric motors, home appliances, and farm machinery. On the other hand, custom-built or low-production equipment and machinery often have rolling element bearings throughout. The latter is a consequence of two situations, namely, the availability of rolling element bearings at low cost, and the reluctance of designers to commit the reliability of their product to a "home-designed" sliding bearing. The rolling element bearing industry developed rapidly as a separate entity as did such products as tires, razor blades, measuring devices, gears, motors, vacuum systems, watches, and tool steels. A particular combination of product precision, distinctive technology, and industry size brought this about. The same did not happen with sliding bearings because these could be made in the machine shops of innumerable industries. Though sliding bearings are easy to make, they are not easy to design. Designers who lack confidence in their ability to design sliding bearings (and associated lubricating hardware), or who do not think such bearings are very good, or who lack confidence in their machine shop will quickly specify rolling element bearings. This in turn allows an economic level of production of a wide range of rolling element bearings.

One of the most widely held reasons for using rolling element bearings in general consumer products is low rolling loss or friction. Well-designed sliding bearings require about 1.3 to 5 times the energy to operate as rolling element bearings at low to moderate speeds, but they have some advantages. In particular, well-lubricated sliding bearings last much longer than do rolling element bearings, and they are stiffer. The rolling element bearing has a limited life because it eventually fails in fatigue. It also deflects considerably under load. A ball bearing with 1-inch bore, and with 10 balls of $^1/_4$-inch diameter and no preload deforms 3×10^{-3} inch with a 10-pound load, and the deformation increases as $W^{2/3}$. (Every 31.62-fold increase in load increases deflection by a factor of 10.) A sleeve bearing with 10^{-3} inch radial clearance becomes very resistant to further deflection after the first 0.7×10^{-4} inch.

ROLLING ELEMENT BEARINGS

Description

Rolling element bearings were developed at an accelerated pace in the 1940s with the development of gas turbines. In jet engines there is a different mix of factors that influence the decision between sleeve bearings and rolling element bearings. Pumping systems for recirculating oil (needed for sleeve bearings) add weight, whereas rolling element bearings can be mist or even vapor lubricated. Rolling element bearings fail eventually no matter how well they are lubricated, but they last longer than do sleeve bearings after the lubricant supply fails.

There are many types of rolling element bearings as may be seen by consulting the sales brochures of bearing makers. Sections through three simpler types are shown in Figure 14.1, a ball bearing, a roller bearing, and a tapered roller bearing. Loading forces, axial and thrust, are shown somewhat in scale. The ball bearing and the roller bearing can carry a small thrust load but a much larger axial load. Ball bearings are also made with races that have deeper grooves for high-thrust loads. The tapered roller bearing can carry a substantially larger thrust load than the others. For high-thrust loads the tapered roller races will also have projections that bear against the ends of the rollers.

Some bearings contain enough rolling elements to abut each other. More often there are fewer rolling elements, and they are separated by a cage or separator as shown in Figure 14.2.

Life and Failure Modes

Rolling element bearings eventually fail either by contact stress fatigue, or by wear. Some wearing occurs because there is always some micro-slip between the rolling pairs (see *Rolling Friction*, Chapter 6). In addition, there is sliding between the cage and rolling elements, sliding of rollers on the races because

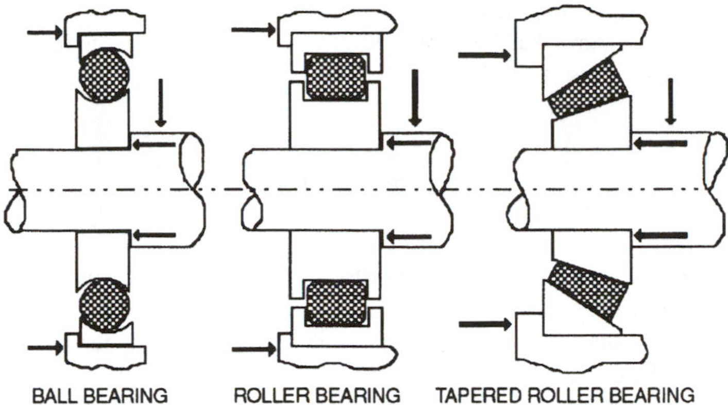

Figure 14.1 Sections of the three most common rolling element bearings. The inner race of the bearing is fitted snugly to the shaft, and the outer race is fitted into a seat in the machine frame or housing.

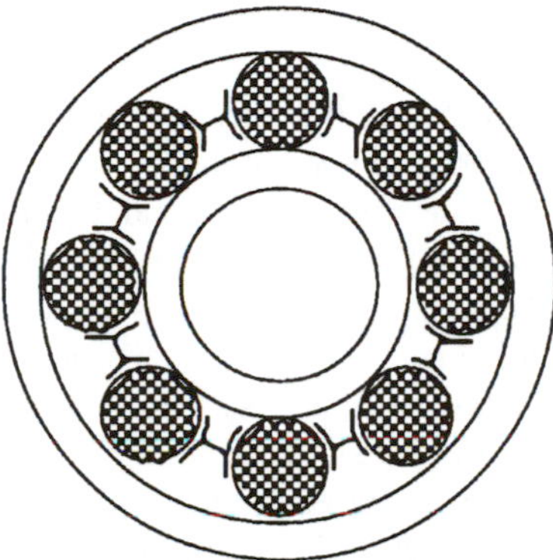

Figure 14.2 Schematic view of a rolling element bearing with roller separators shown.

they prefer to roll along a curved path, sliding of balls in races when the inner and outer races are displaced due to thrust loads, etc. Further, skidding may occur: a rolling element rolls between the races at the loaded side but may lose contact in the unloaded side, stop turning because of friction against the cage, then skid up to speed again as it enters the loaded region. Skidding may be prevented by making the bearing assembly with a slight interference fit between the races and the rolling elements.

Failure of rolling element bearings *by wear* can be prevented by good design, careful manufacture, and proper lubrication. Failure will then inevitably occur by material fatigue. In bearings, flakes of metal spall from the surface of either the roller or from the races, usually the inner race in low speed use (because of the greater counter formal contact) and the outer race in high speed use (due to centrifugal forces).

Bearing manufacturers publish the life of bearings for various applied loads. These data come from well-controlled tests. In general, fatigue life is known to be related to the severity of applied stress. Data from the standard oscillating beam fatigue test show that Nf, the number of cycles to failure, is related to the maximum shear stress, τ, in the metal by:

$$N_f \propto \left(\frac{1}{\tau_{max}}\right)^9$$

This equation is not readily applied to rolling element bearings because of the difference in stress states and difficulty in determining the number of stress cycles that any point in the bearing components experiences per revolution of either race or bearing.

One equation for bearings gives the effect of load on bearing life, L, where P is the equivalent load, n=3 for ball bearings, and C is the dynamic load capacity

$$L_{10} = \left(\frac{C}{P}\right)^n$$

(or the basic load rating), which is the load the bearing can carry for a million inner ring revolutions with 90% chance of survival.

The equivalent load includes two factors, namely, the applied load and the centrifugal loading both multiplied by the appropriate geometric factors of the bearing. This brings up the manner in which the severity of operation of bearings is often expressed in literature relating to high-speed bearings as for turbine bearings. At high speeds the severity is described in terms of DN, where D is the bore diameter in mm and N is the shaft speed in rpm. A bearing of large D has a large number of rollers which, for each turn of the shaft, subjects the bearing race to more cycles of strain than the small-bore bearing would. Jet engines operated in the range of DN between 1.5 and 2 million up to the 1980s. Centrifugal loading can be a significant fraction of the total load. Actually the centrifugal load increases as N^2 such that the severity factor would require a different exponent than 1. An increase in speed from 1.8 to 4.2 million DN reduces the life of a 120 mm bearing by 90% at a load of 2000 pounds and 98% at a load of 4000 pounds.

For more common use the manufacturers' data are adequate. Their tests are usually done at some shaft speed, e.g., 500 rpm. Since fatiguing is a stochastic process, there will be a range of time to failure for a given group or population of bearings. Manufacturers publish the time at which 10% of the population has failed in the form of the B_{10} life, etc. Conservative designers may prefer a B_0 life but this is not available. In response, designers will often select bearings that will carry a much greater static load than their design static load.

In practical use, only about 10% of bearings achieve their expected life. (Many of them are not used to the point of expected life.) Most of those that fail do so because they are poorly made (poor material, cage imbalance or failure, or skidding), some due to misuse (misalignment, shock load, dirt ingestion, or inadequate lubrication), and others because of careless selection.

Load-carrying capacity is directly related to hardness. This usually results in bearings being made of steel of higher than 60 Rc hardness. The tempering temperature of hardened steel limits their operating temperature range to between 400 and 600°F depending on the type of steel. There is also an optimum hardness difference in a bearing: the balls should be 1 or 2 Rc points harder than races. Of secondary importance are the nonmetallic inclusions and trapped gases in steel. If these are reduced by several vacuum meltings, bearing life may increase 4 to 5 times, as occurred during the 1980s. Carbides are also detrimental, but these can be made less harmful by breaking them up during ausforming. With any inclusion, a fiber forms in the ball or race. In the pole regions of the ball (so designated from the practice of hot cropping blanks for forging balls from bar stock), and less so in the equator, fatigue spalls are 10 times more likely to occur if inclusions are present. Compressive residual stress increases bearing life. Pre-nitriding and/or pre-over-stressing doubles bearing life.

Alternate materials: Ceramic bearings are suggested for high temperature service. Alumina, titanium carbide, silicon carbide, and silicon nitride have been used. Homogeneity and porosity are the biggest problems. The best ceramic material available up to 1985 is a cold-pressed alumina which has a C value which is 15% that of M-1 steel.

Lubricating systems: Lubrication is useful to prevent contact of asperities. Viscosity (η) is an important factor: $L \propto \eta^n$ where $0.2 < n < 0.3$. Apparently additives increase effective η at surfaces (see *Boundary Lubrication*, Chapter 9), but such additives as chlorinated wax shorten bearing life by a factor of 7 at worst, by making steel more susceptible to fatigue.

SLIDING BEARINGS

Sliding bearings have many shapes and materials. The simplest shape is the journal (shaft) and sleeve pair as shown in Chapter 9. Thrust loads can also be carried on sliding bearings, but there must be tilting pads to capture lubricant.

Bearings can be made of any material provided the complete separation of the sliding members can be assured. However, in practice, systems must start and stop, they are sometimes overloaded or under-lubricated, dirt gets into them, and they become misaligned. For these purposes, either the system must be redesigned, or material must be selected that accommodates abuse.

The consequences of severe contact conditions must also be accommodated by the choice of material. This choice has two effects. For economy again, crank shafts are often made of nodular cast iron, in which there are graphite nodules on the order of 0.001-inch diameter. Some of these nodules are cut through during grinding, leaving spherical pits, the edges of which often are turned upward. These edges damage bearings.

One of the two sliding surfaces can be made of special materials to extend the conditions for survival of the bearing pair. The four major conditions for survival are:

1. Resistance to fatigue (where there is cyclic loading)
2. Resistance to corrosion (particularly due to acids from combustion)
3. Resistance to scoring (due to inadequate lubrication and high temperature)
4. Ability to embed a limited amount of hard contaminant.

There is no single best bearing material for all types of uses of bearings. Each type of engine, each manufacturing process sequence, each type of oil, each use requires a different bearing. Much experience is required to select the best material.

The manufacturers of bearing materials do specify the broad categories. For example, of the four qualities given above, the lead and babbitt alloys are poorest in resistance to fatigue; the copper-based alloys are the poorest in resistance to corrosion in modern lubricants and with modern fuels; the aluminum alloys are the poorest in resistance to scoring; and the silver and aluminum alloys are poorest in embedding of contaminants.

(See Problem Set question 14.)

MATERIALS FOR SLIDING BEARINGS

An engine bearing is made up of bearing material attached to a steel backing. Layers of different alloys produce galvanic corrosion and some of the elements in the alloys migrate out into other layers. Bearing surfaces may achieve a temperature of 160°C in use. The best overlay material is composed of two phases in which there are either hard particles in a soft matrix, or vice versa for smearing qualities.

The lead and babbitt bearing materials are used mostly in low speed and lightly loaded machinery. Most engines now use alloys based on aluminum or copper. Bearing material must be strong enough to survive, but there is no good way of predicting the needs of bearing materials in terms of measurable properties of the bearing alloy. High (fatigue) strength would be necessary, but the alloys of highest strength have other deficiencies. For example, high-strength materials are less likely to embed debris than are softer alloys. (Ability to embed is to some extent a function of debris particle size and the clearances between the sliding members.) Filters are used to remove most particles over 2 μm diameter.

Bearing alloys are chosen for their low probability of welding to the shaft. Most crankshaft bearing alloys contain a soft, low-melting-point phase which smears over the bearing surface whenever high temperatures are generated in areas of distress. It appears to be best if the smeared metal had not been cold-worked.

Corrosion resistance is needed, particularly where lubricants become very acidic due to long oil-change intervals and short-distance driving. Cavitation can also occur in bearings.

To resist all of the conditions imposed on bearings, it has been found by experiment that a layer of bearing alloy about 0.2 to 0.5 mm thick on a steel backing works well. For more severe applications a third layer of about 0.025 mm thick of soft lead-based alloy is electro-deposited.

Following is a summary of the various types of bearing alloys:[1]

1. *The Lead- and Tin-based Whitemetals (babbitts)*
   - The lead group consists mostly of compositions near: PbSb10Sn6 and PbSb15Sn1As1 which are made up of cuboids of SbSn in a pseudo-eutectic of Pb-Sb-Sn (arsenic refines the Sb precipitate)
   - The tin group is mostly of composition near SnSb8Cu3 which consists of needles of Cu6Sn5 in a SnSb solid solution (Te refines the compound; Cd may be added for strength)
2. *The Copper-Lead Series.* Up to 50% Pb is good for embedding debris and can be operated without an overlay. However, this alloy has poor corrosion and fatigue properties. Lead-free fuel produces more acid in the lubricant than did the former fuel. Modern Cu-Pb alloys have no more than 30% Pb and up to 3% Sn. This is a distinct two-phase structure with Sn concentrating in the Cu when it is present. All are overlay-plated with Pb-Sn, or Pb-Sn-Cu, or Pb-In, mostly to reduce corrosion of the lower layers of alloy. The alloys are CuPb23Sn1 or CuPb30, or for higher loads, CuPb14Sn3, or CuPb23Sn3.
3. *The Aluminum Series.* These alloys need no protection from corrosion. A common alloy is AlSn20Cu1, in which there are connected islands of reticular Sn in an AlCu alloy. Sometimes Sb, Si, Pb, or Cr may be added as well. For some applications AlPb6-8Sn0.5-1.5 with up to 4% Si in some cases, also with traces of Cu, Mg, or Mn for increased fatigue strength.

   For the most severe applications AlSn6NiCu1 or AlSn6Si1.5Ni0.5Cu1, or AlSi4Cd1, or AlCd3Cu1Ni1 or AlSi11Cu1 is used but overlay plated with PbSn or PbSnCu. Small engines might use AlZn5Ni1Pb1Mg1Si1. Aluminum alloys with 12 and 27% Zn are also used.
4. *Bearings for Uses Other than for Crankshaft Bearings.* For many general devices, lubrication is achieved by wick, splash, or mist, and in some instances grease is specified. Wear is a greater problem than corrosion or fatigue in these applications. Most alloys are Cu-based. Polymers may also be used, particularly where there is likely to be poor lubrication. CuPb23Sn3 is used in automatic transmissions, refrigeration compressors, and hydraulic gear pumps. For higher wear resistance use CuPb10Sn10. Solid bronzes are also available, containing CuSn5Pb5Zn5 or CuSn10Pb5 or CuSn10P1. The latter is expensive but stronger than the first two. CuAl8 has been used but it seizes too readily. Whitemetals are usually SnSb8Cu3. SnZn30Cu1 is anodic to steel and thus is useful for marine applications. Small electrical motors use tin-based whitemetals. Acetal copolymer is good, often performing better than bronzes where there is sparse lubrication. Phenolic or polyester resin impregnated into cloth is a good bearing material and works well with water.

   Porous (10–25% pores) bronze is commonly used in bearings for small shafts, where the bronze is impregnated with oil. These cost more but are more effective than molded nylon or acetal resins.

   Dry bearings will tolerate a much wider temperature range than will oil-lubricated bearings and will tolerate vacuum, stop-start, and flat surface sliding. The most popular such bearings are based on PTFE, sometimes impregnated into the bronze, along with some lead. Some bronze bearings contain pockets of graphite and may again contain some lead and tin.

5. *Grooves in Bearing Surfaces*. Pressure lubricated sleeve bearings almost always have grooves of various form, never straight across nor very many, if any, in the heavily loaded areas. The general idea is to bring lubricant to the center of the bearing so that it may flow outward and around the loaded area when load is applied.

REFERENCE

1. Pratt, G., Private communication.

Problem Set

CHAPTER 2

a. Derive Equations 2 and 3, using the notation in Figure 2.4.
b. Plot the Mohr circles for a cube on which there are only equal shear stresses and show how work-hardening progresses (toward brittle failure) as the stresses increase.
c. Show that hydrostatic stresses cannot cause yielding.
d. Show how triaxiality can be beneficial in soft and thin grain boundary films.
e. For a steel having plastic properties described by the equation (in English units), $\sigma = 105{,}000\varepsilon^2$, what is tensile strength?
f. T_g for tire rubber is about – 40°C and its transition occurs over 8 orders of 10 (see Figure 2.13) at constant temperature. When would you expect higher rolling loss of auto tires, in summer or winter?
g. Note the difference in Mohs number for the two crystalline forms of BN in Table 2.4 and explain why there is such a difference.
h. Explain how tensile and compressive residual stresses can form.

CHAPTER 3

a. What is the largest space that could accommodate an interstitial alloy atom such as C or N in BCC and FCC lattice structure?
b. Plot the MP versus E for elements listed in Table 3.2. Are these unique properties?
c. How quickly does water vapor condense on a surface as compared with N_2? Note that a molecule of H_2O occupies 53% more surface area than does a molecule of N_2. ($N_2 \approx 6.2Å^2$)

CHAPTER 4

a. How would grain structure influence the sizes and shapes of laps and folds? (See Figure 4.1)
b. Estimate the ductility of the laps and folds.
c. Measure and estimate the radius of the cutting edge of a steak knife, a razor blade, and a surgical scalpel.

CHAPTER 5

a. Why should a hardness indenter be at least 3 times as hard as the tested surface?
b. What is the source of the force that tears an adhering sphere from a flat surface?
c. What is the real area of contact between a rubber shoe sole and a concrete walk? Assume that the rubber has a 10s modulus of 1.5 GPa. Calculate for running and standing.

d. Plot the temperature at the surface of a ¼-inch-square pin sliding on a flat plate with 50 lb. of load, where μ=0.25. Calculate three cases over a range of VL/2k from 0.05 to 100:

- A copper pin on a titanium plate
- A copper pin on an aluminum plate
- A nickel pin on a manganese plate

CHAPTER 6

a. Plot the HP of energy absorption by brakes when a Lincoln Town Car slows from 60 mph to a stop at a deceleration rate, a′, of 0.5g.
b. Add two columns of data to the table at the bottom of page 77, for $\alpha = 4$ and $\alpha = 16$.
c. Under what conditions might the friction of Nylon 6-6 be less than that of PTFE?
d. How did Roberts and Johnson justify a vanderWaals force of 45 grams?
e. Describe the physics behind Equation 9.
f. How could strain gages on the root of the bar in Figure 6.36 be connected to measure the friction, and to cancel friction?
g. How should strain gages be attached and connected to prevent vertical forces producing some effect on the measurement of horizontal forces?
h. What percentage of practical sliding surfaces function with dead loads, versus spring loads or between "rigid" walls?
i. In Figure 6.36, assume the bar is 1/4″ × 1″ × 10″ and the head is a 1-inch cube (all steel), the resolution of strain gage systems is 10–6 in. and the maximum strain the gages can withstand is 0.002. What is the range of force F that can be measured and what is the primary natural frequency of the cantilever?
j. In Figure 6.38, how much lead distance (see Figure 6.37) would be offset by $\varepsilon = 1°$? Explain.
k. What is the response rate of the typical data acquisition system (including the sensors, amplifiers, etc.), and what variables control this response rate?

CHAPTER 7

a. Measure, the contact angle for water on clean glass, on waxed glass and on various other surfaces.
b. What speed is required to "get up on" water skis, and on bare feet? Is weight a factor?
c. Plot the % of load carried by asperities, or the $A_{real}/A_{apparent}$, for Λ values (see page 165) ranging from 0.001 to 10.
d. Calculate μ between a tire and road for water films of 10^{-9}, 10^{-6}, and 10^{-3} inch thick, with and without molecular effects of viscosity.

CHAPTER 8

a. In Figure 8.1, the CuO and ZnO are not present in the same proportions in the transfer films as the Cu or Zn in the brass. Where is the excess Cu or Zn?
b. Redraw curves in Figure 8.4 to reflect the influence of temperature.
c. Explain the conclusions shown in Figure 8.5.
d. What surface reactions and mechanics could explain the results shown in Figure 8.7?
e. Sketch a device for measuring the strength of attachment of an oxide to its substrate.
f. What could be the magnitude of the error that results from estimating wear rate from the equilibrium values of wear rate at the end of 12 hours in Figure 8.14?
g. What is the maximum hardness of steel that is equally likely to be abraded by both sand and SiC?

CHAPTER 9

a. Does a correlation between the results of an endurance test and a step load test suggest any particular mechanism(s) of wear? (See Figures 9.4 and 9.5)
b. There are several "condition" criteria for scuffing, whereas Figure 9.6 shows a "time" effect. How can "condition" criteria be revised to reflect "time" effects?
c. Describe how to include a "contact shape" factor in scuff criteria.
d. Indicate which "s" curve in Figure 9.11, could apply to the case of loosening and dispersing of lap and fold materials as contaminants in the contact region. Explain.

CHAPTER 10

a. In the equation for the deflection of a cantilever beam ($\delta = PL^3/3EI$). Suppose that the role of L were not yet known. Discuss how the omission of L influences the applicability of the equation. How would you predict δ in some new design?
b. The construction of equation $VT^n f^a d^b = C_1$ is contingent on an exponential relationship between each of "T," "f," and "d" and tool life. Suppose the influence of "f" on tool life were linear, what format of equation would you suggest?
c. Equation 1, suggests a linear relationship between the factors and wear rate. Under severe conditions of sliding one effect of both W and V would be to heat the materials which might alter both "k" and "V." Explain.

CHAPTER 11

a. What is the necessary scale of observation needed to assure that simulation is occurring between a test device and a full-scale machine? Give an example.
b. Of the mechanical properties listed in the center column of the table titled Section D on page 204, how would you measure the bond strength between particles and the surrounding matrix?

CHAPTER 12

Compare the capabilities of both the SEM and metallurgical microscope. That is, what might you see in one and not the other?

CHAPTER 13

List 5 items available in the hardware store in each of the following classifications:

- Surface treated
- Surface modified
- Surface coated

CHAPTER 14

Can sufficient force be transmitted through an oil film to either fatigue or plastically deform bearing materials? Show how you arrived at your conclusion.

Index

A

B

C

D

E

F